T0295356

LASERS AND ELECTRO-OPTICS RESEARCH AND TECHNOLOGY

STATISTICAL MODELS OF CHARACTERISTICS OF METAL VAPOR LASERS

LASERS AND ELECTRO-OPTICS RESEARCH AND TECHNOLOGY

Additional books in this series can be found on Nova's website under the Series tab.

Additional E-books in this series can be found on Nova's website under the E-books tab.

LASERS AND ELECTRO-OPTICS RESEARCH AND TECHNOLOGY

STATISTICAL MODELS OF CHARACTERISTICS OF METAL VAPOR LASERS

SNEZHANA GEORGIEVA GOCHEVA-ILIEVA
AND
ILIYCHO PETKOV ILIEV

Nova Science Publishers, Inc.
New York

For permission to use material from this book please contact us:
Telephone 631-231-7269; Fax 631-231-8175
Web Site: http://www.novapublishers.com

Additional color graphics may be available in the e-book version of this book.

Library of Congress Cataloging-in-Publication Data

Gocheva-Ilieva, Snezhana Georgieva.
Statistical models of characteristics of metal vapor lasers / authors, Snezhana Georgieva Gocheva-Ilieva, Iliycho Petkov Iliev.
p. cm.
Includes bibliographical references and index.
ISBN 978-1-61324-293-3 (hardcover)
1. Metal vapor lasers--Mathematical models. 2.Metal vapor lasers--Design and construction--Statistical methods.I. Iliev, Iliycho Petkov. II. Title.
TA1695.G63 2011
621.36'63--dc22
2011010158

Published by Nova Science Publishers, Inc. † New York

CONTENTS

PREFACE

"Although this may seem a paradox, all exact science is dominated by the idea of approximation."
- Bertrand Russell

Since their invention in 1965, metal vapor lasers (MVL) and in particular metal halide lasers, have been developing extremely rapidly, due to their wide range of technical, medical, ecological and scientific applications. They continue to be the focus of scientific research with over 2000 monographs, books and papers published in renowned journals and presented at conferences.

Nowadays, semiconductor and solid-state lasers dominate over the other types of laser devices because of their excellent technical characteristics - compact size, relatively low price, easy maintenance and long service life. Nevertheless, metal vapor and metal compound vapor lasers continue to be a competitive alternative due to their specific qualities – higher output power and higher laser beam quality [1-3]. Of this group of lasers we have considered the copper and copper compound vapor lasers. Copper and halide compound vapor lasers are among the most powerful laser sources in the visible spectrum (with wavelengths of 510.6 nm and 578.2 nm) and have a high coherence and convergence of the laser beam, among other advantages. In the last few years these lasers were developed as sources of ultraviolet (UV) radiation. Many studies concentrate on lasers with an active volume comprised of copper halide compounds with subsequently added hydrogen or active admixtures: HCl, HBr, etc. These modifications make it possible to improve energy and frequency characteristics, as well as the quality of the laser beam. In this sense, the improvement of existing laser systems and the development of new ones are of great importance. In this context, we will mention some of the newly developed MVLs and laser systems [4-9].

It is considered that MVLs have been studied in relatively good detail both theoretically and experimentally. In addition to the purely experimental studies, there are a large number of analytical and numerical methods, with new ones still being developed, describing various laser systems. Modeling is an irreplaceable and crucial prerequisite for the successful development of lasers. The models are primarily used to solve problems which clarify the experiment and explain the complex physics processes, and all other aspects, which cannot be done through direct measurement. Although big progress has been achieved in that direction, there are still a number of problems which have not been properly solved, both in terms of understanding the processes taking place within the discharge and in a purely constructive respect [1-2]. At the same time, the practical need for the development of new laser devices necessitates a reduction of construction time and costs. Mathematical models evaluating laser variables and output characteristics have continuously played a crucial role in this process. Still, the models constructed so far in the field of metal vapor and metal halide lasers have proven insufficient and incomplete, and statistical analyses have only been carried out within the last few years.

The bulk of the developed mathematical and computer models are the so-called kinetic models, which are based on a mathematical description of the physics processes at work within the discharge, formulated using a large number of coupled differential problems. At the same time, over the years, a large volume of experimental data has been accumulated, which for one reason or another has remained outside of the attention of scientists. However, it is apparent that these data contain summarized and essential information about particular types of laser systems. By systematizing and processing actual experimental data, this information can be extracted and used to study the lasers in more detail.

In this book, with the help of different statistical methods, the available data for two type of metal vapor lasers are used to solve the following problems:

(1) Data cloud dimension reduction;
(2) Classifying input laser characteristics (independent variables) which influence output laser power and laser efficiency;
(3) Establishing multivariate linear, nonlinear and nonparametric regression models of these dependences;

(4) Experiment prediction, based on the models constructed for the purpose of modeling existing laser devices and designing new lasers;
(5) Interpreting the results and clarify the most important physical quantities which influence laser operation.

The statistical analysis is based on experimental data, obtained and published in the last 30 years by the Laboratory of metal vapor laser, Institute of Solid State Physics, Bulgarian Academy of Sciences.

The monograph includes six chapters with bibliography.

Chapter 1 presents a short introduction to metal vapor lasers (MVL) as the main object of study. Their basic characteristics and applications are rendered. A short overview is given of the most important existing papers on mathematical and computer models of MVLs, including kinetic models and models based on experimental data.

Chapter 2 describes in short the basic multivariate statistical methods and techniques, which are later on used for modeling MVLs. There are included cluster and factor analysis for the classification of variables, multiple regression analysis and multivariate adaptive regression splines as a nonparametric technique.

Chapter 3 presents the results from the statistical modeling of the efficiency and power of the copper bromide vapor lasers under consideration. Three main groups of input variables, called factors, have been extracted. These are used to construct linear regression models.

In chapter 4, polynomial and nonlinear models, which improve on the results achieved in chapter 3, are built. The constructed parametric models are compared. It is determined that the polynomial and nonlinear models are of the equal quality.

Chapter 5 presents the methodology and results of the modeling of laser efficiency and power with the help of the MARS technique using all available data. The constructed models are characterized by excellent statistical indices and predict experiment results with accuracy, comparable to that of the experiment itself. Future experiments are predicted.

The last chapter 6 contains results of the classification of the variables and the application of MARS in the case of the available data on a UV CuBr laser. The obtained estimates indicate very good correspondence with experimental data. The quality of the models for experiment prediction is also demonstrated.

The presented models are obtained using SPSS, Wolfram *Mathematica* and SPM MARS software. On request the data files could be sent from snow@uni-plovdiv.bg.

REFERENCES

[1] C. E. Little, Metal vapour lasers: *Physics, engineering and applications*, John Wiley and Sons, Chichester, 1999.

[2] N. V. Sabotinov, *Metal vapor lasers*, In Gas Lasers, eds. M. Endo and R. F. Walter, CRC Press, Boca Raton, 2006, 450-494.

[3] C. E. Webb, *Handbook of Laser Technology and Applications*, 1-3, Institute of Physics Publishing, 2004.

[4] O. S. Andrienko, V. A. Dimaki, G. S. Evtushenko, V. B. Sukhanov, V. O. Troitskiy and D. V. Shiyanov, Metal and metal halide vapor lasers: new opportunities, *Opt. Eng., SPIE*, 44 (2005) 071204.

[5] B. Mao, B. Pan, Y. Wang, L. Chen, and L. Wang, 20-W sealed-off CuBr vapour laser excited with modified Blumlein circuit, *Chin. Opt. Lett.*, 5 (2007), 591-593.

[6] N. Vuchkov, High discharge tube resource of the UV Cu+ Ne-CuBr laser and some applications, In: *New Development in Lasers and Electric-Optics Research,* ed. W. T. Arkin, Nova Science Publishers, New York, 2006, 41-74.

[7] L. Chen, B. L. Pan, Y. J. Wang, K. A. Temelkov and N. K. Vuchkov, He–SrCl2 vapor laser excited by Blumlein discharge circuit, *Opt. Commun.* 282 (2009) 3953–3956.

[8] E. Le Guyadec, P. Nouvel, P. Regnard, A large volume copper vapor +HCI--H2 laser with a high average power, *IEEE J. Quant. Electron.* 41(6) (2005) 879 – 884.

[9] F. A. Gubarev, V. B. Sukhanov, G. S. Evtushenko, V. F. Fedorov and D. V. Shiyanov, CuBr laser excited by a capacitively coupled longitudinal discharge, *IEEE Journal of Quantum Electronics*, 45(2) (2009), 171-177.

ABSTRACT

In this book, various multivariate statistical techniques are applied in order to assess and predict the experiment. There are developed multiple linear and quasilinear regression and other parametric and nonparametric models for modeling laser efficiency and output power. The best models are obtained using nonparametric method of multivariate adaptive regression splines which can account for the local changes in the behavior of the investigated quantities and possess powerful predictive capabilities.

Up until now, such studies have not been conducted in the field of metal vapor laser. The developed empirical models and the results they yield are new and help clarify the relationships between the basic laser input characteristics and the resulting output variables, which is something that cannot be determined using other methods, including the well known kinetic models.

The obtained results can be considered as a first attempt and foundation for the statistical processing of the accumulated experimental data for guiding future experiment in the field of lasers.

Keywords: Parametric model, nonparametric model, multivariate statistical analysis, multivariate adaptive regression splines, metal vapor laser

LIST OF ABBREVIATIONS

MVL	metal vapor laser
UV	ultraviolet
CuBr	copper bromide
OLS	ordinary least-squares (method)
CA	cluster analysis

FA	factor analysis
RA	regression analysis
PCA	principal component analysis
PCR	principal component regression
MARS	multivariate adaptive regression splines
BF	basis function (in MARS method)
GCV	generalized cross validation
Eff	laser efficiency
Pout	output laser power (laser generation)
LPout	gogarithmic transformation of *Pout*

Chapter 1

INTRODUCTION

ABSTRACT

This chapter begins with a brief introduction to the specific object investigated in this book – metal vapor lasers. The basic technical characteristics and the fields of application of copper bromide vapor lasers and UV copper ion lasers are given.

The rest of the chapter presents the tasks of modeling metal vapor lasers. A brief overview of the existing approaches and models is provided. The approaches are divided into 2 groups – structural and phenomenological. In the case of the structural approach, special attention is given to the existing kinetic models, which are among the most widely used for modeling the processes in the laser medium with the help of differential equations. As an alternative to these, the methods based on the phenomenological approach, which are being developed in the last few years, are described. Among them are the methods, applied in this study.

1.1. OBJECT OF STUDY

1.1.1. General Notes

The term "laser" was introduced in 1957 by Gordon Gould [1] and is an acronym of Light Amplification by Stimulated Emission of Radiation. Initially that term was used to describe the optic principle of lasers, but today it designates a large group of devices, which generate visible or non-visible light by stimulating light emission. Depending on the kind of operating medium, lasers are classified into four main types: solid-state, gas, liquid, and semiconductor lasers. Two modes of operation are possible - continuous or pulsed. Lasers differ significantly in structure, size, output power, and other characteristics. More detailed information is available in a wide range of encyclopedias, monographs, etc. (see e.g. [2, 3]).

MVLs are a separate class of gas and plasma lasers, operating at low and medium pressure. These lasers belong to the broader group of pulsed gas lasers. Their main advantages are the high quality of the laser beam and the narrow range of emitted spectral lines. Their operating principle combines quantum-mechanical and thermodynamic processes within the gas discharge mixture, which consists of a large number of particles: electrons, neutral atoms, ions, molecules, and buffer gas. With MVLs, the lasing medium consists of atom vapors of different metals: copper, strontium, gold, manganese, etc., and with some lasers, metal halides - bromides, iodides, etc. - are also added. Gas mixtures or inert gases such as neon, helium etc., are used as buffer gases. It has been proven via experiments that adding small amounts of hydrogen to the active medium can result in an almost two fold increase in laser efficiency [4, 5].

Laser generation has been achieved using Cu, Fe, Mn, Sr, Pb, and Tm halides. The most detailed studies have been conducted on Cu halide lasers, and especially on copper bromide lasers. The particular physical objects of this study are the following types of metal vapor laser:

- Copper bromide vapor laser
- Ultraviolet ion copper bromide vapor laser

1.1.2. Copper Bromide Vapor Laser (CuBr Laser)

The copper bromide vapor laser is an improved version of the pure copper vapor laser. It is the most powerful and effective laser in the visible spectrum, with a highly coherent and converging laser beam. We have studied variations of this laser, developed by the Laboratory of Metal Vapor Lasers, Georgi Nadjakov Institute of Solid State Physics, BAS, Sofia. The first patent of the Laboratory, associated with this type of laser is [6]. The copper bromide vapor laser is among the 12 laser sources which have found wide practical application and which are of commercial interest [4, 5].

CuBr laser development is considered an essential stage in the investigation of copper lasers as a whole.

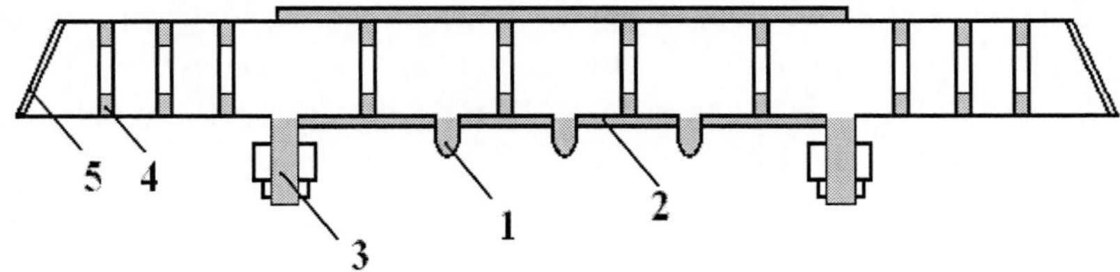

Figure 1.1. Laser tube structure schematic of a copper bromide vapor laser: 1 – copper bromide reservoir, 2 – heat insulation of the active volume, 3– copper electrodes, 4 – internal quartz rings, 5 – mirrors.

Table 1.1. Physical characteristics of a copper bromide vapor laser (see also [5])

Characteristic	Values
Radiation wavelengths	510.6 and 578.2 nm
Operating mode	Periodic pulse, self-heating
Pulse frequency	10-125 kHz
Average volume power density	1.4 - 2 W/ cm^3
Pulse length	20-50 ns
Input power	1 - 5K W
Average output power	1 - 125 W
Coefficient of performance (laser efficiency)	1% - 3%
Total service life	>1000 hours
Pulse energy	6.9 mJ
Active medium temperature	800 - 1000K
Start time	10-15 min
Structural elements	Quartz tube, outer electrodes, copper bromide reservoirs

The copper bromide vapor laser is a well-known pulse radiation source in the visible spectrum (400-720 nm), with two wavelengths: green - 510.6 nm and yellow - 578.2 nm. It is classified as a high-pulsed laser. Neon is used as a buffer gas. Improved efficiency is achieved by adding small quantities of hydrogen. In contrast to the high-temperature pure copper vapor laser, the copper bromide vapor laser is a low temperature one with an operating temperature of the active zone of only 500°C. The laser tube is made out of quartz glass, without any high-temperature ceramics, which makes it significantly cheaper and easier to manufacture. The discharge is heated using electric power (self-heating). Light impulses with a wavelength of several tens of nanoseconds are produced. Its main advantages are: short initial heating period, stable laser generation, relatively long service life, high output power, and laser efficiency. A simplified schematic of the laser is given in Figure 1.1.

The specific technical parameters of the studied copper bromide vapor lasers are given in Table 1.1.

1.1.3. Applications of the CuBr Laser

Copper bromide vapor lasers with various structures and characteristics have a wide range of practical and scientific applications [2, 5]. Figure 1.2 and 1.3 illustrate some of the applications of this type of lasers.

Applications in Medicine and Medical Research

In medicine, the copper bromide vapor laser is mainly used in the fields of dermatology, photocoagulation, cancer treatment, and more. This type of laser is highly suitable for medical applications because of its compact size, short heating period, and the fact that it does not require any pumping. The two lines of the beam are used separately or in combination. The yellow line of laser emission is close to the highest point of oxy-hemoglobin absorption at low melanin mode when treating spots and telangiectasia at a depth of 1-2 mm. The green line is used when treating pigmentations, for specific photo-thermal coagulations of blood vessels, delicate surgical interventions such as palatoplasty, skin rejuvenation, etc. (see [7-11]).

A large number of scientific studies in the field of medicine have been conducted through the practical application of copper bromide vapor lasers (e.g. [12-16] and the cited therein similar papers).

Industrial Applications

Copper bromide vapor lasers and laser systems are used for the micro processing of different kinds of materials: drilling, cutting, marking, engraving, etc. The capability of the laser to achieve precision of up to 10 microns with some materials cannot be matched by any other method. The laser is used to process: metals, plastics, polymers, ceramics, leather, wood, etc. [17-19].

In Scientific Studies

The laser is often used in scientific research: isotope separation of different chemical elements, studies of the magnetic properties of materials, etc. [20, 21].

As a Pumping Source for Other Types of Lasers

As an external power source for maintaining the frequency inversion in the ultraviolet and infrared spectrum, the copper bromide vapor laser is used for pumping titanium-sapphire lasers, dye lasers, etc. [5, 22].

Other Applications

The green-colored beam has the higher gain, and very low attenuation in sea water [23]. Therefore, this laser has applications in the underwater ranging, detection and imaging of the submerged objects like the submarines. Other areas where this laser finds application include forensic use including detection of fingerprints, traces of blood strains, and examination of the documents to check for forgeries [24], pulsed holography, high-speed photography, entertainment and advertising, laser microscopy, laser displays and nanotechnology, military industry, for aerial and naval navigation, in atmosphere and ocean pollution studies, etc. [25, 26].

Figure 1.2. Copper bromide vapor laser, emitting two wavelengths.

Industrial Production

CuBr lasers are manufactured in a number of countries, including the USA, Russia, England, China, India, Bulgaria, the Czech Republic, Israel, Syria, and others.

In particular, licenses for Bulgarian CuBr lasers have been sold to the Australian company Norseld, [11]. It manufactures copper bromide lasers with output power of 10W, 20W, and 40W for industrial and medical applications. In Bulgaria, CuBr lasers are produced by a number of small and medium-sized companies. The most prospective is Pulslight, [27], which specializes in the manufacture of copper bromide vapor lasers (with output power of 2W, 4W, 5W, 10W, 40W, 80W) for laser display units, medical systems and industrial laser systems for precision processing of materials. Pulslight exports its products to the USA, Mexico, Canada, Germany, Poland, Italy, Cyprus, Romania, and other countries. Other Bulgarian companies, which produce lasers and laser systems, based on the CuBr laser are: Laser Product, Spektronika, Optela – Laser Technologies, IILIT "Satura" JSC, ZMM Metalik JSC, Factory for production of industrial and medical lasers, High-technology Park "Hebar" PLC.

Figure 1.3. Laser show.

1.1.4. Ultraviolet Copper Ion Excited Copper Bromide Vapor Laser (UV C+ CuBr Laser)

The copper ion excited copper bromide vapor laser is a promising innovative product with unique characteristics and stable operation. The first lasers of this type were constructed by the Laboratory of MVLs at ISSP of BAS in 1999. This laser emits five different wavelengths in the ultraviolet spectrum. It is characterized by high quality of the laser beam and high output power. Its fields of application include medicine, microelectronics, photolithography, genetic engineering, etc. During the last decade, this type of laser has been the subject of intense experimental studies and its characteristics have been significantly improved [28, 29]. The laser is soon to be at a stage of its development, which allows it to be used in industry.

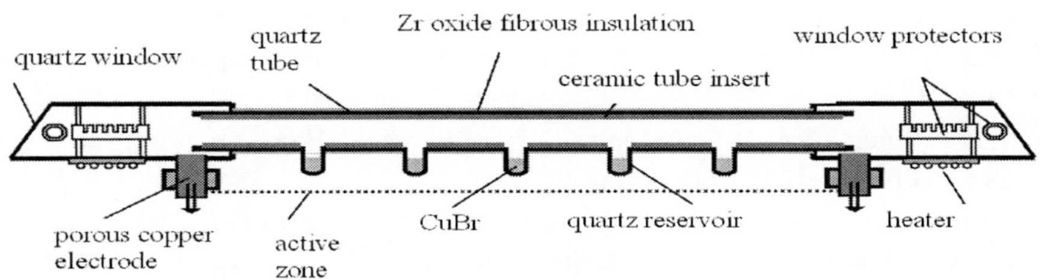

Figure 1.4. Laser tube structure of a UV copper ion excited copper bromide vapor laser.

Table 1.2. Physical characteristics of a UV copper ion excited copper bromide vapor laser [5, 28]

Characteristic	Values
Radiation wavelength	248.6, 252.9, 259.1, 260.0, 270.3 nm
Operating mode	Periodic pulse, self-heating
Pulse rate frequency (PRF)	15-25 kHz
Pulse length	20-50 ns
Input electrical power	1 - 2.5 KW
Average output power	0.5 - 1.3 W
Coefficient of performance (laser efficiency)	1%
Total service life	700 hours
Active medium temperature	1000 K
Start time	15 - 20 min
Buffer gas Neon	7 - 30 torr
Hydrogen admixture	0 - 0.06 torr
Active zone length	60 – 100 cm
Tube diameter	4 – 26 mm
Structural elements	Outer quartz tube, inner ceramic tube insert, outer electrodes, copper bromide reservoirs

Ultraviolet laser generation can be achieved using copper or gold ions. The typical technological problems of copper ion lasers – copper deposits on the inside walls of the tube and high discharge temperature – were solved thanks to the development of a new Ne-CuBr laser with a nanosecond pulsed longitudinal discharge. This laser emits in the deep ultraviolet spectrum on five spectral lines - 248.6, 252.9, 259.7, 260.0, and 270.3 nm. Experimentally, a record average output power of 1.3W has been obtained at all five lines and 0.85W at the 248.6 nm line. It has also been determined that adding small amounts of hydrogen (0.02-0.04 Torr) results in a two-fold increase of laser output power.

The schematic of the ultraviolet laser tube is shown in Figure 1.4. and technical characteristics are given in Table 1.2.

Laser generation has been also achieved with an ultraviolet gold ion laser, which is beyond the scope of this investigation.

1.1.5. Applications of the UV Copper Ion Laser

Due to the narrow emission range of just a few spectral lines and the high coherence of the beam of the copper ion excited copper bromide laser, it is used for processing which requires high-resolution, such as recording information, fluorescence, high-precision drilling, cutting, cleaning, modification of newly developed materials, etc.

Micro Processing of Materials

Highly efficient photochemical ablation is achieved using an ultraviolet copper ion excited CuBr laser in order to produce clean precision cutting and drilling with minimum thermal or mechanical damage to the target. In particular, this allows the processing of polymers, organic material, semiconductors, glass, etc. [4, 28, 30].

Other Applications

In studies and modification of some thin polymer films [31, 32], in medicine, microbiology and Raman-spectroscopy, for pumping of ion dye lasers, etc. [2, 5, 28].

Industrial Production

The biggest and best known company is the American enterprise Photon Systems [33], established in 1997. It produces UV ion copper bromide vapor lasers in four wavelengths for a wide range of technical and scientific research applications.

1.2. Overview of Existing Methods and Models

1.2.1. Structural and Phenomenological Approach and Types of Models

The state of a given system is determined by a certain number of parameters which describe its properties to some degree. The objective of a mathematical modeling is to find the relationships (equations) between the main parameters, which explain the influence of the parameters on the appearance of the properties that are of interest. Usually, a complex model includes a number of simpler models.

Two main approaches are utilized when constructing models of a given system:

- Structural
- Phenomenological.

The structural approach tries to construct a model structure of the medium. In the case of the laser medium, it contains a large number of particles - electrons, positive and negative ions, neutral atoms. Laser output characteristics are considered to be the summary result of the internal movement and interactions of these particles. Mathematically this is described using specific model systems of differential equations and appropriate initial and boundary-value conditions. Obtained models are kinetic and examine the complex physical processes in the active laser medium. Another type of modeling involves simulating the discharge and the processes within, using the so-called large or super particles, the Monte Carlo method, or other methods.

The phenomenological approach does not take into account the structure of the laser medium. Experimental data alone are used to model the practical problem. Most frequently, the results are processed using statistical methods – factor analysis, regression analysis, cluster analysis, etc. Up until now, this approach had not been used in the field of metal vapor lasers, including copper vapor and copper compound lasers. On one hand, this is partially due to the lack of sufficient and complete experimental data, which as mentioned above requires time-consuming and costly experiments. On the other hand, one of the main reasons is the fact that researchers and engineers, working in this field, are not acquainted with appropriate statistical methods for data analysis and experiment planning. It is well known that scientists and engineers need to possess good statistical knowledge. However, an edifying example is the disastrous end of the American space shuttle Challenger in 1986, when a simple initial processing of the data, known before the launch, could have prevented the catastrophe from happening (see [34], page 480).

A successful phenomenological model could be directly linked with the implementation and evaluation of the system, which is being examined and could be used to guide, assist, and analyze experiments when designing new laser sources with improved output characteristics. Some of the main tasks at different stages of the investigations are:

- Filter the entire set of parameters (variables or factors) in order to distinguish the most significant ones, which are to be studied further
- Obtain the reliable dependence based on data and finding the corresponding regression or other appropriate statistical models
- Planning experiments, including the extreme experiment, the main purpose of which is to optimize particular characteristics or the object of the study itself.
- Conducting a design of experiment [35], however very cost for laser planning.

The predominant portion of existing mathematical and computer models of gas lasers operating at low pressure have been developed using the structural approach.

The main types of structural models can be systemized in the following groups [36]:

- Kinetic, describing particle movement
- Fluid (uninterrupted), concerned mainly with the longitudinal changes along the axes
- Simulation models which utilize super-particle methods, including the particle-in-cell, cloud in cell or CIC – method
- Two-group models (plasma is treated as a population of bulk and light particles).

In the field of metal vapor lasers, almost all models are kinetic, with a few exceptions. Publications dealing with the other methods are also significant in number and can be overviewed in the large number of articles and monographs [36-39].

1.2.2. Survey of Kinetic Models

Low temperature plasma (LTP) is a self-consistent system – its physical condition, which determines the chemical activity, depends on the speed of the reactions. Kinetic models are the most frequently used method to qualitatively describe the processes within the pulse discharge at medium and low pressures. The operating medium of the discharge consists of electrons, atoms, and ions, which move and interact, transferring energy through various kinds of collisions: electron – atom, electron – electron, electron – ion. A simplified scheme of energy transformation in a LTP is given in Figure 1.5.

The kinetic model for the plasma uses: the distribution function for electron energies from Boltzmann's kinetic equation, continuity and movement equations, electric field potential and intensity equations, hydrodynamic electron energy equations taking into account the dependence of transport coefficients on electron temperature, neutral gas temperature and electric field intensity, energy balance equation, gas temperature distribution equation, etc. [36-38].

Constructing a kinetic model of a specific laser system or subsystem comes down to determining known and unknown parameters, system equations needed to describe the phenomenon, and the respective boundary and/or initial conditions of the problem. As a rule, a large part of equations are first order stiff ordinary differential equations and/or partial differential equations. In principle, solving the model means combining analytical and numerical methods using a computer.

Because kinetic models are not the subject of this study, here we will briefly describe only the most important results in that direction.

The summarized models are applicable to the following MVLs: pure copper vapor lasers, copper bromide vapor lasers, copper chloride vapor lasers, UV ion lasers, and the so called "kinetic enhanced" (with modified kinetics) copper lasers, developed in the last ten years through adding hydrogen and halides. In particular, pure copper vapor lasers and kinetic enhanced lasers are the subject of numerous models. These lasers have similar characteristics to the ones we are investigating, namely: radiation wavelengths similar to those of a CuBr laser (510.6, 578.2 nm), high output power more than 100W, and high efficiency at about 1%, very good coherence and laser beam quality. We have to note that due to the high operating temperature of the copper laser (around $1600\,^{o}C$) it is characterized by abrupt changes of some laser characteristics in the inter-pulse period, as well as a longer initial discharge induction period, which is one of the main objects of modeling for this type of laser.

The first computer kinetic models in the field of MVLs were developed between 1973 and 1980 at the same time in the USA and in Russia. Later, groups began work in Australia, England, Israel, China, and in other countries.

The main results, achieved using kinetic models have been summarized in review papers [36-44]. The results from more than 300 articles have been surveyed.

The monograph [36], presents in detail the physics and practical side of the experiment, together with the theory and numerical modeling of discharge processes in high-frequency capacitive discharges at low pressure. Numerous models and examples of modeling have been developed.

Paper [37] is an important survey article, describing basic models in the considered field and the state of existing problems. It deals with all aspects of modeling of low temperature plasma: differential equations, (including Boltzmann's equations, Boltzmann's moment equations, fluid approach, Poisson's electric field equation, etc.), numerical methods for solving the equations, numerical modeling of direct (dc) and radio-frequency frequency (rf) discharge, and other currently existing models. The author concludes that "numerical modeling is assured of a long and exciting future."

It is considered that the classics in the field of modeling MVLs are the results of Prof. M. Kushner (now head of the *Computational Plasma Science and Engineering Group* at the University of Michigan) and his research associates. Paper [45] describes a self-consistent computer model of a copper vapor laser, consisting of a system of differential equations with 17 basis variables. It is concluded that due to the limits imposed by the materials used in the construction, copper lasers and in general MVLs operate by far under their potential in terms of laser output power and laser efficiency. Paper [46] studies the kinetics of several copper vapor lasers. The model is based on seven energy levels of copper and neon, and the equations of the electric field, electron temperature (with Maxwell's equation), particle density, electron density, and the distribution of gas temperature T_g.

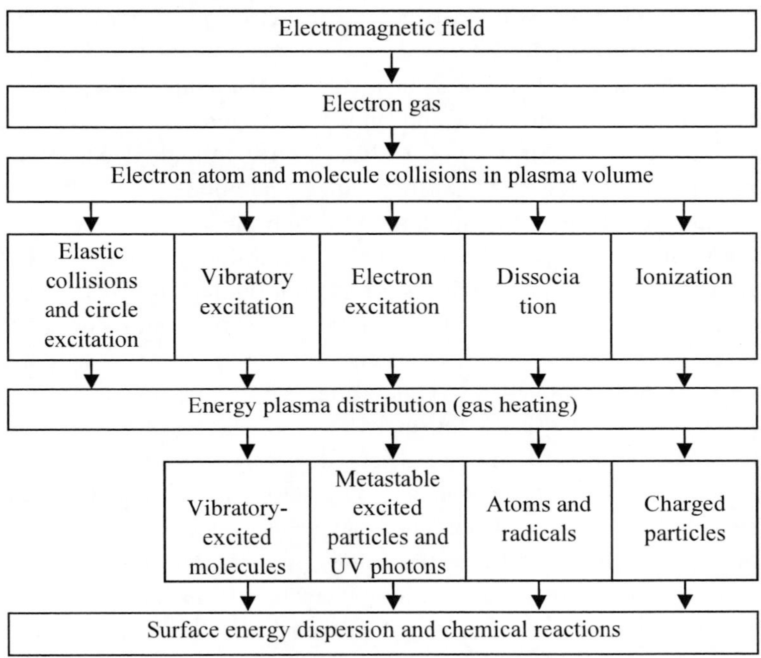

Figure 1.5. A sketch of energy transformation in low-temperature plasmas.

Prof. R. Carman and his colleagues at the *Centre for Lasers and Applications, Macquarie University*, Sydney, Australia have been especially prolific in the field of kinetic modeling of copper lasers and copper halide lasers. In essence, their models are an improvement of Kushner's kinetic models. The article [47] presents a self-consistent kinetic model of a copper vapor laser, which includes 70 collision and emission processes. Various optimal and non-optimal operating conditions for the lasers constructed at their laboratory have been simulated. A very complex self-consistent kinetic model of the gas discharge in a copper vapor laser with tube diameter of 2.5 cm has been described in [48]. The model includes 150 processes for 30 types of particles. The rest of the computer models constructed and developed by Carman's group for various copper lasers and their kinetic enhanced variants examine: the intensity of the electric field and the influence of the electric characteristics, pre-pulse electron density ([49] and other).

A large group of research scientists from *Zhejiang University, Hangzhou*, China has been working for over 20 years in the field of modeling of copper vapor and kinetic enhanced copper lasers, and strontium ion lasers. We will mention the article [50], which describes CuBr laser plasma with five levels using in part experimental data on electron temperature, density, and other parameters. The paper [51] deals with calculating the temperature of the buffer gas of a CuBr laser with neon and added hydrogen. In the article [52] a computer kinetic model of a UV Cu+ laser in a longitudinal pulse Ne-CuBr discharge is constructed for the first time. The model consists of seven ordinary differential equations and examines the basic kinetic processes.

Using a numerical model, [53] studies the influence of the main input parameters on output characteristics of an ultraviolet Ne-CuBr laser. The authors of the article [54] present a parametric optimization of an ultraviolet Ne-CuBr laser using a self-consistent kinetic model. The consistent characteristics are obtained by comparing with experiment results. The influence of tube diameter is established.

The recent article by Iranian scientists [55] presents a mathematical model, describing the kinetic and laser characteristics of a CuBr laser with neon as a buffer gas and added hydrogen. The model offers a standard evaluation of the changes in current, voltage, population of levels (5 energy levels of copper atoms and 3 energy levels of neon atoms), electron temperature, and radial distribution of pressure and the temperature of the buffer gas. The obtained system consists of 13 ordinary differential equations and partial differential equations for the gas temperature.

Articles about MVLs by Russian authors are the most numerous. Their achievements are presented in detail in a special edition of [38]. They mainly deal with copper vapor lasers. These lasers have found wide industrial application in Russia. What is more, there are a number of studies of copper bromide vapor lasers, strontium atom lasers and UV lasers.

The most active in the last few years group of Russian scientists in the field of modeling copper vapor lasers with added halides works at the Tomsk State University. They have published over 40 articles on the subject. Some of their results have been included in doctorate dissertations (see for instance [44]). Interesting results have been obtained in [56, 57], where kinetic models of a copper vapor laser and a copper vapor laser with added hydrogen are presented. In the article [58] using the previous model, the mechanisms connected with the influence of hydrogen at different pulse frequencies have been studied.

The two recent publications [59, 60] present a simulation model to study the generation characteristics of a CuBr laser. In the first article, new kinetic equations are obtained on the basis of precisely perfected particle interactions. The second article supplements the constructed model with new reactions. A total of 240 kinetic reactions have been obtained and these have been used to develop a self-consistent model.

We will also survey the articles by Bulgarian authors in the field of structural modeling of MVLs.

Paper [61] presents the experiment results for a CuBr laser, a model for calculating gas temperature, and the results of the calculations. The studies in [62] are devoted to the influence of the diameter of the active zone on the characteristics of a ultraviolet ion excited copper bromide vapor laser with neon as a buffer gas and frequency of 19.5-25 kHz. In addition to obtained experiment results, a simple kinetic model has been applied in order to describe discharge processes during the initial self-heating phase of the discharge.

During the last few years, in [63-68] new analytical models have been developed in order to improve Kushner's model from 1983 [46], widely used up until now for evaluating radial gas temperature in metal vapor lasers. The models are based on analytic solutions of the steady-state heat conduction equation subject to mixed boundary conditions for the arbitrary form of the volume power density in the internal laser tube, combined with nonlinear boundary value conditions in the remaining part of the composite tube. Explicit analytical formulas for calculating the radial temperature profile of the laser tube have been obtained, which take into account all heat exchange processes, as well as the conditions of the surrounding media. They can be applied without providing experimental values of the wall temperatures, and could be used for designing new laser tubes. The models have been developed in details for different types of lasers, especially for copper bromide vapor lasers, ultraviolet ion copper lasers, and high-powered strontium bromide lasers (He-$SrBr_2$). Various computer simulations have been carried out in order to evaluate engineering solutions, which utilize the models. Good coincidence has been shown between experiment results and other simpler models.

Numerical models for the potential and intensity of the electric field in the longitudinal section of the tube in copper bromide vapor lasers during one laser impulse have been developed and solved by means of a special type of finite difference scheme in [69].

We also have to note some results from the studies in the field of computer implementation and application of the models we created in order to develop specialized software. In accordance with this, the capabilities of the BPEL language for design and management of scientific processes, in combination with various simulation modules (servers) as web-services were applied to the copper bromide vapor lasers in question. Technology was applied which allows the packing of ready-made executable code for numerical simulations which allows us to solve constructed analytical models [70-72].

In addition, a working version of the prototype of LasSim (Laser Simulation) software was recently developed (see [73]).The prototype is designed on the newest .NET Framework, WPF and C# technologies. It gives the opportunities for installing and managing the model simulation processes and carrying out the simulations in metal vapor lasers using developed mathematical models in [63-69].

1.2.3. Survey of Known MVL Models Constructed Using the Phenomenological Approach

As mentioned above, new type of models can be obtained when a phenomenological approach is applied. Such models have only begun to be used in the field of MVLs in the last few years and have yet to receive the necessary recognition from the physics community due to the reasons pointed out in this chapter.

An interesting result was obtained in [74], where the statistical method of the orthonormal design of the experiment was applied in order to optimize the parameters of the discharge electric circuit of a CuBr laser. It is reported that an optimal set of parameters for effective and stable operation of the laser was obtained.

In the publications [75-77], different genetic algorithms have been developed in order to optimize the inductive-capacitive parameters of the electro-discharge circuit and to study the kinetic processes in the plasma of a kinetic enhanced (optimized) copper vapor laser. Paper [78] presents a genetic algorithm for modeling a copper vapor laser with added hydrogen. The inductive-capacitive properties of the discharge circuit and its laser impedance are examined.

In the general overview of the publications, we will also note some of the results in the field of statistical modeling of MVLs, which we have recently obtained.

Based on experimental results accumulated during the last 30 years at ISSP of BAS, statistical processing has been conducted using different multivariate statistical models. A number of models were constructed with the aim to describe the influence of laser parameters on output laser characteristics such as efficiency and power. It was determined that in the case of copper bromide vapor lasers, 12 basic input laser parameters can be classified in 3, 4 or 5 groups according to different agglomerative methods of cluster analysis. The influence of the groups on laser generation and efficiency was also determined [79, 80]. The development of a procedure for making predictions based on cluster and factor analysis was started in [80]. Multidimensional factor and regression analyses were applied in [81, 82] for a copper bromide vapor laser, using various factor rotation methods and different regression analysis methods. For different samples of 6 input parameters (tube length, tube diameter, diameter of the rings, applied electric power, hydrogen pressure, relative power per volume unit), the last were grouped in 3 factors. Using explicit formulas, linear regression models which describe the dependence of laser efficiency and generation on the factors were obtained. An appropriate nonlinear regression model was obtained in [83]. In [84, 85] multivariate adaptive regression splines are used for obtaining nonparametric and more flexible models for copper bromide and ultraviolet ion excited copper bromide lasers.

The basic approaches, methods, and results from our statistical studies are described later on in the chapters that follow.

REFERENCES

[1] G. R. Gould, The LASER, Light Amplification by Stimulated Emission of Radiation, In *Proc. of The Ann Arbor Conference on Optical Pumping*, the University of Michigan, 15 June through 18 June 1959, eds. P. A. Franken and R. H. Sands, 1959, 128.

[2] C. E. Webb, *Handbook of Laser Technology and Applications*, 1-3, Institute of Physics Publishing, 2004.

[3] O. Svelto, Principles of lasers, 3rd ed., Plenum Press, New York, 1998.

[4] C. E. Little, Metal vapour lasers: *Physics, engineering and applications*, John Wiley and Sons, Chichester, 1999.

[5] N. V. Sabotinov, *Metal vapor lasers*, In Gas Lasers, eds. M. Endo and R. F. Walter, CRC Press, Boca Raton, 2006, 450-494.

[6] N. V. Sabotinov, P. K. Telbizov and S. D. Kalchev, Bulgarian patent N 28674, 1975. http://www.issp.bas.bg/lab/LabMVLs/.

[7] http://veinlase.com/treatveinlas.htm.

[8] http://www.smartskincare.com/treatments/noninvasive/nonablativelaserlight_copper-bromide_511-578nm.html.

[9] http://dermnetnz.org/procedures/copper-bromide-laser.html.

[10] http://www.aestheticsurgery.com.au/Content_Common/pg-sydney-vascular-laser-treatment- specialists.seo - Sydney Vascular Laser Skin Treatments by Copper bromide laser.

[11] http://www.norseld.com/.

[12] S. McCoy, M. Hanna, P. Anderson, G. McLennan and M. Repacholi, An evaluation of the copper–bromide laser for treating telangiectasia, *Dermatol. Surg.* 22(6) (1996) 551-557.

[13] S. Loginov, Sokolova, G. N., E.P. Markin and R. V. Ambartsumyan, Using copper-vapor lasers to treat stomach ulcers, *Bull. Acad. Sci. USSR Phys. Ser.* 54(10), (1990) 70-79.

[14] D. A. Cassuto, Non-ablative photorejuvenation with a scanned copper bromide. Laser, Lasers in Surgery and Medicine, *ASLMS* (2003) 54.

[15] O. Sabotinov, E. Stoykova, *Copper-bromide laser system for treatment of dermatological malformations*, Proc. SPIE, eds: P. A. Atanasov, S. V. Gateva, L. A. Avramov and A. A. Serafetinides, Bellingham, WA, 5830 (2005) 449-453.

[16] L. Longo, M. G. Postiglione, O. Marangoni and M. Melato, *Two-year follow-up results of copper bromide laser treatment of striae*, J Clinical laser medicine and Surgery, 21(3) (2003) 157-160.

[17] P. G. Foster, *Industrial applications of copper bromide laser technology*, Thesis (Ph.D.), University of Adelaide, School of Chemistry and Physics, Dept. of Physics and Mathematical Physics, 2005.

[18] I. Balchev, N. I. Minkovski, N. V. Sabotinov and I. K. Kostadinov, Micromachining with copper bromide laser, Proc. SPIE, eds: P. A. Atanasov, A. A. Serafetinides and I. N. Kolev, Bellingham, WA , 5226 (2003) 372-376.

[19] P. L. G. Ventzek, R. M. Gilgenbach, C. H. Ching, R. A. Lindley, W. B. McColl, *Copper vapor laser machining of polyimide and poly-methylmethacrylate in atmospheric pressure air*, *J. Appl. Phys.* 72(7) (1992, 2009) 3080 – 3083.

[20] B. Warner, *Atomic vapour laser isotope separation*, Proc. SPIE, 737 (1998) 2–6.

[21] S. E. Roozmeh, M. M. Tehranchi, M. Ghanaatshoa, S. M. Mohseni, M. Parhizkari, H. Ghomi and H. Latifi, *Magnetoimpedance effect in laser annealed* $Co_{68.25}Fe_{4.5}Si_{12.25}B_{15}$ *amorphous ribbons*, *J. of Magnetism and Magnetic Materials,* 304(2) (2006) e633-e635.

[22] Z. Lei, Q. Liejia, Z. Guiyan, S. Xiudong and L. Fucheng, *Generation of 30ps pulse from a short cavity dye laser pumped by a copper bromide laser*, Chinese Phys. Lett. **7** (1990) 300-303.

[23] T. Reghunath, V. Venkataraman, D. V. Suviseshamuthu, R. Krishnamohan, B. R. Prasad, S. Raghuveer, C. K. Subramanian, P. Chandrasekhar and P. S. Narayanan, *The origin of blue-green window and the propagation of radiation in ocean waters, Def. Sci. J.* , 41(1) (1991) 1-20.

[24] J. Hecht, *The laser guide book*, McGraw-Hill Book Co., New York, 1986, 163-173.

[25] G. S. Evtushenko, V. Yu. Kashaev, N. V. Parshina, V. B. Sukhanov and V. V. Tatur, *Light-dynamic systems based on copper bromide vapor laser*, Proc. SPIE 4900 (2002) 1126-1129.

[26] X. Zureng, Z. Guiyan and L. Fucheng, *Applications of the CuBr vapor laser as an image-brightness amplifier in high-speed photography and photomicrography*, Appl. Opt. 31 (1992) 3395-3397.

[27] http://www.pulslight.com/.

[28] N. Vuchkov, High discharge tube resource of the UV Cu+ Ne-CuBr laser and some applications, In: *New Development in Lasers and Electric-Optics Research,* ed. W. T. Arkin, Nova Science Publishers, New York, 2006, 41-74.

[29] N. K. Vuchkov, K. A. Temelkov and N. V. Sabotinov, *UV laser system for materials processing*, *The Journal of the Bulgarian Academy of Sciences,* 1 (2006) 39-41.

[30] E. Little, Metal vapour ion lasers: *Kinetic processes and gas discharges*, John Wiley and Sons, New York,1996.

[31] K. Beev, K. Temelkov, N. Vuchkov, Tz. Petrova, V. Dragostinova, R. Stoycheva-Topalova, S. Sainov and N. Sabotinov, Optical *properties of polymer films for near UV recording*, *J. Optoelectr. Adv. Materials*, **7** (2005) 1315-1318.

[32] M. Ilieva, V. Tsakova, N. Vuchkov, K. Temelkov and N. Sabotinov, *UV copper ion laser treatment of poly-3,4- ethylenedioxythiophene*, J. Optoelectr. Adv. Materials, 9 (2007) 303-306.

[33] http://www.internatlaser.com/laser_systems/copper_vapor_laser_systems.html, *Laser Show Systems For Advertising, Entertainment And Attractions,* accessed: 25 Dec 2010.

[34] C. Montgomery and G. C. Runger, *Applied statistics and probability for engineers,* 3rd edition, John Wiley and Sons, New York, 2003.

[35] R.A. Fisher, *The Design of Experiments*, Oliver and Boyd, Edinburgh, 1935.

[36] Yu. P. Raizer, M. N. Schneider and N. A. Yatsenko, *Radio-frequency capacitive cischarges*, CRC, New York, 1995.

[37] G. G. Lister, *Low-pressure gas discharge modeling, J. Phys. D: Appl. Phys*. 25 (1992) 649-1680.

[38] *Encyclopedia of low-temperature plasma, Series B, vol. 7: Numerical modeling of low-temperature plasmas*, Ed. M. Ianus, Moscow, 2004 (in Russian).

[39] Bogaerts, K. De Bleecker, V. Georgieva, D. Herrebout, I. Kolev, M. Madani and E. Neyts, *Numerical modeling for a better understanding of gas discharge plasmas, High Temperature Material Processes*, 9(3) (2005) 321-344.

[40] J. van Dijk, G. M. W. Kroesen and A. Bogaerts, *Plasma modelling and numerical simulation*, Review article, *J. Phys. D: Appl. Phys*. 42 (2009) 190301.

[41] G. G. Petrash, *The processes limiting the pulse repetition rate in pulsed metal and metal compound vapor lasers, Laser Physics*, Springer 10(5) (2000) 994–1008.

[42] R. J. Carman, *Computer modelling of longitudinal excited elemental copper vapour lasers*, In *Pulsed metal vapor lasers – physics and emerging applications in industry, medicine and science*, eds. C.E. Little and N.V. Sabotinov, Kluwer Academic Publishers, Dordreht, 1996, 203-214.

[43] M. J. Withford, D. J. W.Brown, R. P. Mildren, R. J. Carman, G. D. Marshall and J. A. Piper, *Advances in copper laser technology: kinetic enhancement, Progress in Quantum Electronics*, 28(3-4) (2004) 165-196.

[44] O.V. Zhdaneev, *Modeling laser processes in copper bromide lases with kinetic enhancement*, PhD thesis, Tomsk, Russian Academy of Sciences, 2004 (in Russian).

[45] M. J. Kushner, *A self-consistent model for high repetition rate copper vapor lasers*, J. *Quant. Electr*. 17 (1981) 1555-1565.

[46] M. J. Kushner and B. E. Warner, *Large bore copper vapor lasers: Kinetics and scaling issues, J. Appl. Phys*. 54(6) (1983) 2970-2982.

[47] R. J. Carman, D. J. W. Brown and J. A. Piper, *A self-consistent model for the discharge kinetics in a high-repetition-rate copper-vapor laser, IEEE J. Quant. Electron*. 30(8) (1994) 1876-1895.

[48] R. J. Carman, R. P. Mildren, M. J. Withford, D. J. W. Brown and J. A. Piper, *Modeling the plasma kinetics in a kinetically enhanced copper vapor laser utilizing HCl+H2 admixtures, IEEE J. Quant. Electron.* 36(4) (2000) 438-449.

[49] M. J. Withford, Investigations of the effect of trace impurities on copper vapour laser performance, *PhD Dissertation*, Macquarie University, 1995.

[50] Cheng and W. Sun, A kinetics model and study of CuBr pulsed lasers, *Acta Physica Sinica*, 41(10) (1992) 1605-1612.

[51] T. Ma, *Analysis of gas temperature radial distribution in Ne-H2-CuBr laser discharge tube with diaphragms*, Yingyong guangxue (*Journal of Applied Optics*), 27(1) (2006) 54-57.

[52] B. L. Pan, G. Chen, B. N. Mao and Z. X. Yao, *Kinetic process of UV Cu+ laser in Ne-CuBr longitudinal pulsed discharge*, *Opt. Express*, 14 (2006) 8644-8653.

[53] B. N. Mao, G. Chen, Y. B. Wang, L. Chen and B. L. Pan, *Numerical study of the influence of the parameters on laser operation in UV Ne-CuBr laser in longitudinal impulse discharge*, Wuli xuebao, 56(6) (2007) 2652-2656.

[54] B. N. Mao, B. L. Pan, L. Chen, Y. J. Wang and Z. X. Yao, *Kinetic analysis of the factors limiting the output power of the Ne–CuBr UV laser*, *Chinese Phys*. B, 18 (2009) 1542-1546.

[55] B. A. Ghani and M. Hammadi, *Modeling the plasma kinetics mechanisms of CuBr laser with neon-hydrogen additives*, *Opt. and Las. Technol.* 38 (2006) 67-76.

[56] M. Boichenko, G. S. Evtushenko, S. I. Yakovlenko and O. V. Zhdaneev, The influence of the initial density of metastable states and electron density on the pulse repetition rate in copper-vapor laser, *Laser Physics,* Springer, 11(5) (2001) 580-588.

[57] M. Boichenko, G. S. Evtushenko, O. V. Zhdaneev and S. I. Yakovlenko, *Theoretical analysis of the mechanisms of influence of hydrogen additions on the emission parameters of a copper vapour laser, Quant. Electron.* 33(12) (2003) 1047-1058.

[58] M. Boichenko, G. S. Evtushenko, S. I. Yakovlenko, O. V. Zhdaneev, *Theoretical analysis of mechanisms behind the influence of hydrogen admixtures on lasing characteristics of a copper-vapor laser, Laser Physics,* Springer, 13(10) (2003) 1231-1255.

[59] M. Boichenko and G. S. Evtushenko, *Simulation of a CuBr laser in the presence and in the absence of hydrogen impurity, Laser Physics,* Springer, 18(4) (2008) 403–412.

[60] M. Boichenko, G. S. Evtushenko and S. N. Torgaev, *Simulation of a CuBr laser, Laser Physics,* Springer, (12) 18 (2008) 1522-1525.

[61] D. N. Astadjov, N. K.Vuchkov and N. V. Sabotinov, *Parametric study of the CuBr laser with hidrogen additives, IEEE J. of Quant. Electr.* 24(9) (1988) 1926-1935.

[62] N. K. Vuchkov, K. A. Temelkov, P. V. Zahariev and N. V. Sabotinov, *Optimization of a UV Cu+ laser excited by pulse- longitudinal Ne-CuBr discharge, IEEE J. Quantum Electron.* 37(4) (2001) 511-517.

[63] P. Iliev, S. G. Gocheva-Ilieva and N. V. Sabotinov, *Analytic study of the temperature profile in a copper bromide laser, Quantum Electron.* 38(4) (2008) 338-342.

[64] P.Iliev, S. G.Gocheva-Ilieva and N. V.Sabotinov, *An improved model of gas temperature in a copper vapour laser, Quantum Electron.* 39(5) (2009) 425-430.

[65] P. Iliev, S. G. Gocheva-Ilieva, K. A. Temelkov, N. K. Vuchkov and N. V. Sabotinov, *Modeling of the radial heat flow and cooling processes in a deep ultraviolet Cu+ Ne-CuBr laser,* Mathematical Problems in Engineering, Hindawi Publishing Corporation, 2009, Article ID 582732, 17 pages.

[66] P. Iliev, S. G. Gocheva-Ilieva, *Model of the radial gas-temperature distribution in a copper bromide vapour laser, Quantum Electron.*, 40(6) (2010) pp. 479 – 483.

[67] P. Iliev, S. G. Gocheva-Ilieva, K. A. Temelkov, N. K. Vuchkov and N. V. Sabotinov, *Analytical model of temperature profile for a He-SrBr2 laser, J. Optoelectron. Adv. Mat.* 11(11) (2009) 1735 - 1742.

[68] P. Iliev, S. G. Gocheva-Ilieva, K. A. Temelkov, N. K. Vuchkov and N. V. Sabotinov, *An improved radial temperature model of a high-powered He-SrBr2 laser, J. Opt. Laser Technol.* Elsevier 43(3) (2011) 642-647.

[69] S. G. Gocheva-Ilieva and I. P. Iliev, *Mathematical modeling of the electric field in copper bromide laser,* Proc. Int. Conf. of Numerical Analysis and Applied Mathematics, ICNAAM 2007, September 16-20, 2007, Corfu, Greece, American Institute of Physics (AIP), CP936 (2007) 527-530.

[70] Malinova and S. Gocheva-Ilieva, *Application of the business process execution language for building scientific processes for simulation of*

metal vapour lasers, Conf. Proc. 3th Balkan Conf. in Informatics (BCI'2007), 27-29 September 2007, Sofia, Bulgaria, 2 (2007) 75-86.

[71] Malinova, *Software system for computer simulation of metal vapor lasers*, PhD thesis, Plovdiv University Press, 2009.

[72] Malinova, *Design approaches to wrapping native legacy codes*, Scientific works of Plovdiv University 36(3- Mathematics) (2009) 89-100.

[73] S. G. Gocheva, C. P. Kulin, *Development of LasSim software prototype for simulating physical characteristics of laser devices*, Scientific Works of Plovdiv University, 37(3-Mathematics) (2010) 45-52.

[74] J. L. Lu and L. J. Wang, *The orthonormal design of experiments for the optimization of the parameters of the discharge circuit in the CuBr vapour lasers power supply, Laser Technology*, 30(2) (2006) 113-115.

[75] Cheng and S. He, *An optimal design for reducing the black center for a copper-vapor laser by using a genetic algorithm, Microwave Opt. Technol. Lett.* 25(2) (2000) 113-119.

[76] Cheng, *Plasma kinetics mechanisms of an optimized copper vapour laser, J. Phys. D: Appl. Phys.*, 33 (2000) 1169-1178.

[77] Cheng and S. He, *On the optimization of laser power, efficiency and impedance matching for a copper vapor laser, Microwave Opt. Technol. Lett.* 27 (2000) 339-343.

[78] Cheng and S. L. He, *Optimal design for a copper vapor laser with a maximum output by using a genetic algorithm, Optical and Quant. Electron.* 33 (2001) 83-98.

[79] P. Iliev, S. G. Gocheva-Ilieva and N. V. Sabotinov, *Classification analysis of the variables in a CuBr laser, Quantum Electron.* 39(2) (2009) 143-146.

[80] P. Iliev, S. G. Gocheva-Ilieva and N. V. Sabotinov, *Prognosis of the Copper Bromide Laser Generation through Statistical Methods*, in XVII Intern. Symposium on Gas Flow and Chemical Lasers and High Power Lasers 2008, September 15-19, Lisbon, Portugal, eds. R. Vidal et al., Proc. of SPIE, Bellingham, WA 7131, 2009, 71311J1-J8.

[81] P. Iliev, S. G. Gocheva-Ilieva, D. N. Astadjov, N. P. Denev and N. V. Sabotinov, *Statistical approach in planning experiments with a copper bromide vapor laser*, Quantum Electron. 38(5) (2008) 436-440.

[82] P. Iliev, S. G. Gocheva-Ilieva, D. N. Astadjov, N. P. Denev and N. V. Sabotinov, *Statistical analysis of the CuBr laser efficiency improvement*, Opt. Laser Technol. 40(4) (2008) 641-646.

[83] S. G. Gocheva-Ilieva and I. P. Iliev, *Nonlinear regression model of copper bromide laser generation*, Proc. COMPSTAT'2010, eds. Y. Lechevallier, G. Saporta, 19th Int. Conf. Comp. Statistics, Paris - France, August 22-27, Springer_ebook, *Physica-Verlag*, 2010 , 1063-1070.

[84] S. G. Gocheva-Ilieva and I. P. Iliev, *Parametric and nonparametric empirical regression models of copper bromide laser generation*, Math. Probl. Eng., Theory, Methods and Applications, Hindawi Publ. Corp., New York, NY, Article ID 697687 (2010), 15 pages.

[85] S. G. Gocheva-Ilieva and I. P. Iliev, *Modeling and prediction of laser generation in UV copper bromide laser via MARS*, in *Advanced research in physics and engineering*, series "Mathematics and Computers in Science and Engineering", ed. O. Martin et al., Proc. 5th *Int. Conf. Opt., Astrophysics and Astronomy* (ICOAA '10), Cambridge, UK, February 20-22, 2010, WSEAS Press, 2010, 166-171.

Chapter 2

DATA AND STATISTICAL METHODS

ABSTRACT

This chapter considers the basic approaches and aspects of creating statistical models for application to natural and engineering sciences. The concept of a model and the different types of models are discussed. The specific characteristics of parametric and non-parametric models are clarified. Experimental data for the basic laser characteristics of copper bromide vapor lasers and ultraviolet ion excited copper bromide vapor lasers, which are the subject of further investigation, have been presented. The rest of this chapter contains brief, systematized descriptions of the multivariate statistical methods, used to process the data. These methods are: cluster analysis, factor analysis, regression analysis, principal component regression and multivariate adaptive regression splines.

2.1. SOME ASPECTS OF STATISTICAL MODELING IN PHYSICS AND ENGINEERING

The development and operation of a laser device requires detailed knowledge of dependences between its basis characteristics: its geometric proportions, levels of energy values – input power, halide concentration, etc. Studying the structure of these relationships allows for a more detailed understanding of the physics processes, making it possible to control the increase or decrease of laser output characteristics and other specific results associated with the overall laser operation. The main objectives of this study are concerned with the direction of the experiment, the technology for production, computer simulations of respective processes, predicting the experiment, etc.

There are two basic types of approaches for statistical modeling: global parametric and local nonparametric. Global parametric methods apply strategies that utilize global variables, which are part of statistical models and analyses throughout the intervals or regions being studied and in this way, their influence is taken into account as "a whole", i.e. globally. This strategy is not always appropriate, if it is supposed that the variables taken into account interact in a non-homogeneous manner within their intervals. On the other hand, global techniques are suitable for relatively small datasets, since separate observations have a greater relative influence on the elements of the model.

Local nonparametric methods are used to determine the subregions of input variables, in which a given dependent variable responds in a particular way. Local dependencies are determined for each individual region. In this way, the model is much more flexible and the specifics of the investigated function are described more accurately.

In this paper, the relatively new nonparametric method of multivariate adaptive regression splines (MARS) is used alongside classic parametric methods such as cluster analysis, factor analysis, and principal component regression in order to construct empirical models.

These methods are used to find and study the general dependencies between basis laser characteristics (hereafter referred to as variables) of the lasers, being considered based on existing experimental data.

2.1.1. The Need for Statistical Data Processing

Statistics deals with the collection, representation, analysis and application of numerical data. It assists in solving different practical problems, such as planning and construction of new products and systems, makes it possible to determine the quantitative and qualitative relationships between cause and effect, allows for new laws to be established, etc. During the accumulation and processing of experimental data, statistical processing is necessary in order to facilitate data comparison, experiment planning, and quality control during industrial manufacture, and other activities. More specifically, statistics and statistical techniques serve as a powerful tool, assisting scientists and engineers in their daily work.

Usually it is considered that in the field of natural sciences and engineering the processes being studied are strictly determined. The existence of a large number of laws describing different processes creates the impression that every experiment is more or less a confirmation of a given law. In reality, things are quite different. Especially in more complex systems, where a great number of interactions occur simultaneously (each described by a different law), the influence of one or another input variable of the system on the resulting (output) variables cannot be described precisely. What is more, usually when conducting an experiment, one and the same measurement done under the same set of conditions may produce different results, sometimes this difference is significant. This can be explained by the fact that:

- Each physics, chemistry or other type of law has been formulated to some degree using assumptions and simplifications, which are not always clearly formulated and/or are difficult to adhere to
- Measurement devices are not sufficiently accurate and stable adding an inherent measurement error
- The relative measurement error of different devices for different variables is not the same
- In more complex systems, some processes are highly changeable or unstable and dependent on either time or another process, or a set of processes
- The phenomenon being studied is dependent on processes, which have a weak or corrective effect on measurements, but they are unknown or have not been examined in sufficient detail.

For these reasons, when using data to describe an object or a phenomenon, there is always a degree of variability, which can be subjected to statistical investigation.

2.1.2. Types of Statistical Studies

A set of all possible data, describing the object of study, forms the population. The sample is a subset of a given population. Data in natural sciences, including in physics and engineering, are typically quantitative data or interval (continuous) type. This distinguishes these fields from social sciences, where nominal and ordinal data are also used.

In some cases, the sample is taken from a well-defined and known population. Using samples, conclusions can be drawn for the population as a whole. Each sample introduces the so-called sampling error. The sample has to be of adequate size and quality in order to minimize the sampling error as much as possible.

In most cases, the population is of conceptual type, which allows for measurements to be taken later on. For example, we have only a sample of a certain number of measurements, which can be considered to be representative of the whole population that could be obtained in the future. In these cases, an additional assumption is made that the sources of variability of the properties of future measurements would be the same as those for the sample.

According to the type of data in the sample, a given statistical study may be classified as [1]:

- Retrospective
- Observation
- Planned experiment

Retrospective studies use only the so-called historical data, which are available up to that moment or for a given period. This is necessary in the cases where for one reason or another obtaining new measurement data is either difficult or impossible. Either a sample or all available data are used. The main disadvantage is that these data have usually been collected without having in mind the purpose of a later statistical study, which means that it might be incomplete and some variables may not have been varied sufficiently or they could even have been fixed. For this reason, the results might be unsatisfactory from a practical point-of-view.

During observation, researchers can: follow closely the measurements, monitor and register the accuracy of the data, select the variables that are of interest as well as other elements of data collection and compilation, in accordance with the goal of the statistical study.

Planned experiments play a very important role in practice. They were first proposed by Fisher [2] for crop field planning in agriculture but nowadays they are a well-developed statistical apparatus applicable in various fields. With this type of experiment, researchers can freely choose and monitor the way the experiment is conducted, while at the same time maintaining the strict succession required for statistical analysis. By monitoring output results, researchers are able to draw specific conclusions as to which variables change the most and have the strongest influence on the rest of the variables. There are numerous publications on this subject (see for instance [1, 3-4]).

All of the processed data in this monograph are historical. This is due to the fact that conducting experiments is a very expensive and time-consuming process. This limits the possibilities for statistical studies by means of observations and planned factor experiments.

2.1.3. Parametric and Nonparametric Methods for Studying Dependencies

Let the dependence of a given variable *y*, which is being measured, to be written in the following generalized form:

$$y = f(x_1, x_2, ..., x_p) + \varepsilon \tag{2.1}$$

Where *f* is a deterministic component, dependent on one or more explanatory variables (predictors) $x_1, x_2, ..., x_p$, and ε is an error which influences the measurement but cannot be determined or controlled. Supposedly, we have at our disposal tabular data $\{y_i, x_{1i}, x_{2i}, ..., x_{pi}\}_{i=1}^{n}$. The goal of the statistical modeling is to find a function (or a calculation rule), based on the data in the form

$$\hat{y} = \hat{f}(x_1, x_2, ..., x_p) \tag{2.2}$$

or for the observations

$$\hat{y}_i = \hat{f}(x_{i1}, x_{i2}, ..., x_{ip}),\ i = 1, 2, ..., n\ . \tag{2.3}$$

One expects $\hat{y}$ to fit the data sufficiently well satisfying certain statistical conditions. In (2.2) and (2.3), $\hat{f}$ denotes the empirical function, in which the measurements from the experiments are substituted, and the elements of the vector $\hat{y}$ are the respective calculated values of the dependent variable. The latter is different from *y*.

An equation of type (2.2) or (2.3) is a typical example of a model of variable *y*. It has to be noted that different models can be constructed for one and the same data.

When looking for function $\hat{f}$ dependent on a certain number of parameters $a_0, a_1, ..., a_m$, we talk about parametric modeling, i.e. the model is in the following form:

$$\hat{y} = \hat{f}(x_1, x_2, ..., x_p, a_0, a_1, ..., a_m). \qquad (2.4)$$

If the dependence in this equation is linear with respect to the parameters, model (2.4) is a *linear parametric* one, and if it is nonlinear the model is *nonlinear parametric.* We have to specify that it is assumed that the random component ε has to be independent of $x_1, x_2, ..., x_p, a_1, ..., a_m$ and subject to a probability distribution, usually normal distribution.

When the deterministic component f in (2.1) is known from a physics law or another type of law in the form of a formula, the model is called *mechanistic.* A simple example of a mechanistic model is the integral form of Om's law for a part of an electric circuit:

$$I = \frac{U}{R},$$

where I is the amperage, U is the voltage or the difference in electric potential, and R is the resistance.

Of course, there is always an error in the measurements themselves, as previously discussed. Another example that can be provided is with a known approximate form of the sought dependence, for example, a linear combination of exponential functions $f = \beta_0 + \beta_1 e^{\lambda_1 x_1} + \beta_2 e^{\lambda_2 x_2} + \varepsilon$ and the parameters of the model $\beta_0, \beta_1, \beta_2, \lambda_1, \lambda_2$ have to be found so that known data can be fitted in the best possible way.

Many examples of parametric models from the fields of physics and engineering can be found in [1, 3-6] and others.

However, for more complex systems, the type of dependence (2.1) is completely unknown. Then, constructed models are called *empirical.* They can be concretely represented, the objective being to provide a sufficiently accurate approximation and description of experimental data.

In many studies, data cannot be described by one single function in the form of (2.4) and finding a parametric model is impossible or the developed model is unsatisfactory. Most often, this is due to unfulfilled requirements of statistical analyses, and usually, when the distribution of data differs significantly from the normal or other form of data distribution. Non-parametric models are applied in that case.

With nonparametric models, the dependence (2.1) is evaluated directly instead of looking for unique parameters for the entire multidimensional region of predicators. Therefore, while in classic regression analysis the goal is to estimate the parameters of the model, in nonparametric regression, the goal is to estimate the regression function directly.

It is implicitly assumed that function f is continuous. The main approach is to divide the definition domain of the function into sub-regions and to construct partially approximating functions for each sub-region thus allowing a more flexible approximation and prediction of the dependent variable.

From a statistical point-of-view, each model should satisfy certain hypotheses and criteria for adequacy and validity of its conclusions. In all cases, it is recommended that simpler models are constructed first, for example linear ones. The next step is to look for nonlinear models, if the linear ones are not sufficiently satisfactory. It must be noted that if a model is over-determined there is a very real danger of overestimation, when the model begins to also approximate the measurement error, i.e. when reading the stochastic component ε.

It is important to note that as a whole, the modern approach in statistical modeling is distinguished from the conventional one in that it is much more dependent on the data, than on the choice and experience of the researcher. Simply put, it is much more data than user driven. Since the 80s, modern methods and means for data processing are heavily computer intensive, combining intelligent methods with large volume calculations. They can be used to model complex nonlinear relationships between inputs and outputs or to find patterns in data. The most characteristic methods of this type are based on artificial neural networks (ANN), genetic algorithms, classification trees (CART) ([7, 8]) etc.

Some modern methods such as, Generalized Additive Models [9] utilize a given set of target variables and a set of candidate predictors to determine the type of descriptive function according to the data. Other methods (CART, MARS) determine on their own both the significant predictors and the functional form of the dependence [10, 11].

2.2. Data Description

2.2.1. Data for a Copper Bromide Vapor Laser

Statistical analyses of CuBr lasers are conducted based on experimental results, published in [12-22].

A total of over 270 experiments have been examined.

This includes different types of copper bromide vapor lasers with various input and output characteristics. The general type of the geometry and laser characteristics are given in Chapter 1. In literature, the CuBr lasers are divided into three main groups according to their geometric characteristics:

- Small-bore lasers, with an inside diameter of the tube, id $\leq$ 20 mm,
- Medium-bore lasers, with id in the interval (20,40] mm
- Large-bore lasers - with id more than 40 mm.

The investigated 8 types of lasers and the respective references are given in Table 2.1. The short notation of the type id40/20x50 has been assumed, designating in this case a 40 mm inside diameter of the tube, 20mm inside diameter of the rings and 50cm active length.

Table 2.1. Types of CuBr lasers

Type	Reference
id15 /4.5x30	[14, 22]
id40/20x50	[12, 15, 20, 22]
id40/40x50	[18, 20]
id40/40x120	[18, 22]
id46/20x50	[12, 13]
id50/30x100	[21, 16]
id50/30x140	[21, 16]
id58/58x200	[17, 19, 22]

The data for 12 observed variables, crucial for the operation of the laser system, are studied. Of them, 10 are independent and 2 are dependent. They are described in detail in Table 2.2.

Table 2.2. Copper bromide vapor laser characteristics under investigation

Independent input variables	Unit
D – inside diameter of the laser tube	mm
DR – inside diameter of the rings	mm
L – distance between the electrodes (active zone length)	cm
PIN – input electrical power	kW
PH2 – hydrogen pressure	Torr
PL – power per unit length	kW/cm
PRF – pulse repetition frequency	kHz
PNE – neon pressure	Torr
C – equivalent capacity of the condenser battery	nF
TR – temperature of the copper bromide reservoir	^{0}C
Dependent output variables	
Pout – output laser power (laser generation)	W
Eff – laser efficiency	%

2.2.2. Data for Ultraviolet Ion Excited Copper Bromide Laser

We use experimental data for the ultraviolet laser, obtained in the last few years, the main part of which has been published in [23-27]. Some of the data have been generously provided by collaborators at the Laboratory of Metal Vapor Lasers at the Institute of Solid State Physics, Bulgarian Academy of Sciences, Sofia.

A total of 251experiments have been examined.

The data for observed basic variables, crucial for the operation of the UV laser, as described in Table 2.3, are studied.

The quantity *TR*- temperature of the copper bromide reservoir – is not considered to be a variable, since as a result of experimental studies, its optimal value has been determined to be $TR = 560\,^{0}C$.

Table 2.3. Investigated characteristics for a UV copper ion excited CuBr laser

Independent input variables	Unit
D – inside diameter of the laser tube	mm
L – distance between the electrodes (active zone length)	cm
PIN – input electrical power	kW
PV – power per unit volume	kW/cm^3
PH2 – hydrogen pressure	Torr
PL – power per unit length	kW/cm
PRF – pulse repetition frequency	kHz
PNE– neon pressure	Torr
TR – temperature of the copper bromide reservoir (fixed optimal value)	560°C
Dependent output variables	
Pout – output laser power (laser generation)	W
Eff – laser efficiency	%

2.3. Multivariate Parametric Methods Used to Construct Empirical Models

We will describe in brief the main methods and assumptions, which serve as the basis of our statistical study.

2.3.1. Methods for Initial Data Processing and Transformations of Data

1) Notations and Data Type

All of the data used for statistical processing are measured at the interval and ratio level. The observed quantities, which are random in nature, are to be called input or independent variables. Their respective values are to be denoted by the rectangular matrix

$$\mathbf{X} = \begin{pmatrix} x_{11} & x_{12} & \cdots & x_{1p} \\ x_{21} & x_{22} & \cdots & x_{2p} \\ \cdots & & & \\ x_{n1} & x_{n2} & \cdots & x_{np} \end{pmatrix} \quad (2.5)$$

This matrix can be given in the following two concise forms:

$$\mathbf{X} = (x_{ij}), \; i = 1,...,n, \; j = 1,...p \quad (2.6)$$

$$\mathbf{X} = \begin{pmatrix} x_1' \\ x_2' \\ \vdots \\ x_n' \end{pmatrix} = \left(x_1, x_2, \ldots, x_p\right),$$

where x_j is the column vector of the *j*-th variable and x_i' is the row vector, comprised of the i -th respective values of each variable x_j.

In some statistical analyses, we will also use the matrix:

$$\tilde{\mathbf{X}} = \begin{pmatrix} 1 & x_{11} & x_{12} & \cdots & x_{1p} \\ 1 & x_{21} & x_{22} & \cdots & x_{2p} \\ \cdots & & & & \\ 1 & x_{n1} & x_{n2} & \cdots & x_{np} \end{pmatrix} \quad (2.7)$$

When studying dependencies from data **X**, we will denote the cases of the dependent variable using the vector

$$y = (y_1, y_2, ..., y_n)', \quad (2.8)$$

where $'$ denotes the transpose.

It is considered that dependent variables are not random in nature. Our goal will be to determine if a dependance does actually exist (primarily of a correlative type); what is its strength and form.

2) Distribution Type

Usually, when conducting most statistical techniques using multi-dimensional interval data for parametric models, it is required that the data are subject to multivariate normal distribution or that the distribution does not deviate significantly from the normal one. Techniques, such as cluster and factor analysis, do not necessarily require multivariate normality [28]. In our case, from a physics point-of-view, we assume that the set of variables for a copper bromide laser are almost normally distributed. However, the actual obtained data have been measured for independent variable values, chosen by the researcher, so in the general case are not completely random.

Nonparametric regression methods, also called free distribution methods, are applied in the cases, where the conditions for multivariate normality have not been met and the size of the sample is not very big. We will use this method as well, in order to construct nonparametric models.

3) Samples

There are numerous methods for selecting different types of statistical samples: simple random sample, cluster sample, random proportional sample, stratified sample, etc. As mentioned, in our case experiments have not been conducted in an entirely random way, which necessitates the selection of samples, providing an improvement of their randomness. Random samples cause the smallest additional error. The particular simple random samples from the investigations have been obtained using the corresponding algorithm of the SPSS software package. If the selected sample meets the requirements of a given statistical technique, it is considered to be consistent and can be used to conduct analyses. Conclusions about the general set are drawn on the basis of its statistics.

We have to note that it is also possible to perform bootstrap (bootstrap - sampling from the sample) procedures in order to define obtained statistics [29]. The latter is an extremely computer-intensive processing method during which artificial samples are generated (possible case repetitions) and all 500-1000 such samples are analyzed, which is not among the objectives of this study.

4) Transformation of Data into a Standardized Form

Due to the difference in the physical units of the data variables, usually for most of the multivariate techniques, dimensionless reduction is performed beforehand, i.e. the so-called standardized form of the data is obtained though a transformation with a mean 0 and standard deviation 1. When applied to the data for a given random variable $t=(t_1,t_2,...,t_n)$, this is achieved by calculating its z –values using the following formula:

$$z_k = \frac{t_k - \overline{t}}{s}, \quad k = 1,2,...,n, \tag{2.9}$$

where n is sample size, $\overline{t}$ - its mean, and s is the standard deviation of the sample. This transformation preserves the distribution of the data.

5) Box-Cox Power Transformation

Some special variance-stabilizing transformations can be used in order to improve the normality and the symmetry of the distributions, as well as their approximation to a multivariate normal distribution [30].

In case of strictly positive variables x, the Box-Cox transformation is well-known [31]:

$$\psi_{BC}(\lambda, x) = \begin{cases} (x^{\lambda} - 1)/\lambda, & \lambda \neq 0 \\ \log(x), & \lambda = 0 \end{cases} \tag{2.10}$$

where usually the parameter $\lambda \in [-2,2]$.

6) Yeo-Johnson Power Transformation

In 2000, in their paper [32] Yeo-Johnson published the following generalization of the Box-Cox transformation (2.10) for an arbitrary sign of the variables:

$$\psi_{YJ}(\lambda, x) = \begin{cases} \left\{(x+1)^{\lambda} - 1\right\}/\lambda, & x \geq 0, \lambda \neq 0 \\ \log(x+1), & x \geq 0, \lambda = 0 \\ -\left\{(-x+1)^{2-\lambda} - 1\right\}/(2-\lambda), & x < 0, \lambda \neq 2 \\ -\log(-x+1), & x < 0, \lambda = 2 \end{cases} \tag{2.11}$$

The Yeo-Johnson transformation has a number of advantages, including continuous first and second derivatives with respect to λ, $\lambda \in [-2,2]$. Both independent and dependent variables can be transformed [32]. It has been proven that this and the previous transformation improve the mutual multivariate distribution of the data.

2.3.2. Cluster Analysis

Cluster analysis (CA) is the common name of a number of calculation procedures, used to classify objects. It makes it possible to find new dependencies and characteristics of the laser system, which cannot be determined using other known theoretical or experimental methods.

2.3.2.1. General Characteristics

In cluster analysis, classified objects can be both observations and variables. The concept of a cluster can be considered as universal. Generally speaking, a cluster is a subset (group) of objects which are similar to a certain degree, according to certain criteria.

As a reduction technique, CA shares elements with discriminant and factor analysis. For example, with discriminant analysis units are also classified but in predefined groups. The cognitive approach for the two types of analysis is also different. Discriminant analysis investigates the differences between known groups and the possibilities for developing an effective rule for unit classification, while CA investigates the possibilities to set up logically founded and practically explainable uniform groups (clusters) of units. FA classifies variables in general factors and not units.

The goal of CA is to find an optimal grouping of observations or variables, for which the elements of a given cluster are similar but the clusters are clearly distinguishable from each other.

Usually with CA, the number of groups is not known in advance and is determined by the researcher during the investigation.

There are different types of CA, the most widely used are: hierarchical methods and the K-means method (partitioned or non-hierarchical). In the case of interval data and a relatively small sample size (under 500), hierarchical methods are more suitable and we will describe them in more detail.

Formed groups (clusters) need to be similar (homogenous) within and different (heterogeneous) from each other according to specified characteristics. The quantitative evaluation of the term "similarity" is connected with the term "metric”. With this similarity approach, events are represented as points in a given space. Determined similarities and differences between the points are defined in accordance with the metric distances between them.

Two objects are identical, if the distance between them is equal to zero. The greater the distance between the two objects, the greater the difference between them.

A very detailed description of a large number of techniques, methods and the use of SPSS for CA has been given in [33-36, 8].

2.3.2.2. Basic Requirements for Applying Cluster Analysis

CA has no specific discriminating requirements towards data, its distribution and interdependencies. There is no mechanism for discerning variables which are suitable or unsuitable for the investigation. If some significant variables, are ignored, results may be inadequate. The other problem is finding an optimal number of clusters, which is not a formalized procedure.

When the solution is correctly defined, in the case of CA according to variables, many authors consider that the results should be formally close to those of FA for the same data.

2.3.2.3. Measures of Similarity or Dissimilarity

There are many distance functions that can be used to measure proximity (similarity) between each pair of observations or variables. When the distance between a given pair of objects is increasing, it is considered to be a measure of dissimilarity.

1) Euclidean Distance

The most frequently used measure of distance between interval data is Euclidean distance. For two vectors $x = (x_1, x_2, \dots, x_p)'$ and $y = (y_1, y_2, \dots, y_p)'$ in $\mathbb{R}^p$ (where $'$ denotes the transpose) it is defined as:

$$d(x, y) = \sqrt{(x-y)'(x-y)} = \sqrt{\sum_{k=1}^{p} (x_k - y_k)^2} \qquad (2.12)$$

2) Squared Euclidean Distance

$$d(x,y) = \sum_{k=1}^{p} (x_k - y_k)^2 \tag{2.13}$$

3) Chebyshev Distance

$$d(x,y) = \max_{k=\overline{1,p}} |x_k - y_k|$$

4) Minkowski Distance of Order *r*

$$d(x,y) = \left(\sum_{k=1}^{p} |x_k - y_k|^r \right)^{1/r}$$

This way, in the case of n observations or variables we have to calculate the proximity matrix $\mathbf{D} = d(x_1, x_2)$, which is symmetric and has zero elements along the main diagonal.

For example, let us take three objects (points in a two-dimensional space) with measurements for two variables (x_1, x_2): (4,7), (4,2), (2,3). Here $p = 2, n = 3$. Using the usual Euclidean distance, we calculate the distances:

$$d_{12} = \sqrt{(4-4)^2 + (7-2)^2} = \sqrt{25} = 5$$

$$d_{13} = \sqrt{(4-2)^2 + (7-3)^2} = \sqrt{20} \approx 4.5$$

and

$$d_{23} = \sqrt{(4-2)^2 + (2-3)^2} = \sqrt{5} \approx 2.2$$

For the distance matrix we get

$$\mathbf{D_1} = d(x_1, x_2) = \begin{pmatrix} 0 & 5 & 4.5 \\ 5 & 0 & 2.2 \\ 4.5 & 2.2 & 0 \end{pmatrix}$$

However, it is obvious that the distance depends on the measuring unit. For example, if we multiply the first variable x_1 by 10 (assuming we convert centimeters to millimeters), the proximity matrix would change to

$$\mathbf{D_2} = d(x_1, x_2) = \begin{pmatrix} 0 & 5 & 20 \\ 5 & 0 & 20 \\ 20 & 20 & 0 \end{pmatrix}$$

Now the lowest distance is d_{12}, instead of the previous d_{23}.

This problem is usually solved using preliminary data standardization through z -scores or another method.

Other kinds of distances can also be used. For example, a widely used measure is the Manhattan distance or city block distance, and others.

2.3.2.4. Clustering Methods

After choosing a measure of similarity or dissimilarity between objects and the distance matrix is calculated, we move on to the second stage of CA – clustering. The numerous methods, which have been developed in statistical literature to this end, can be divided into two groups: hierarchical (vertical) clustering methods and non-hierarchical (horizontal) clustering [33, 34].

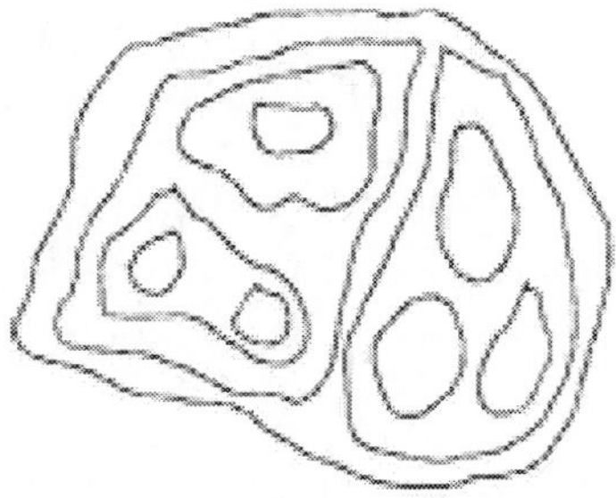

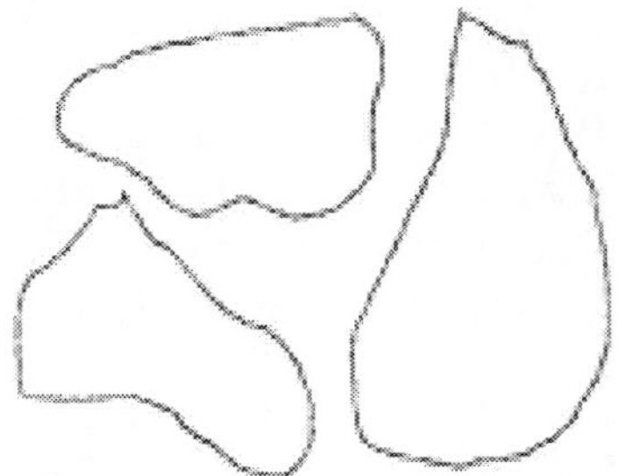

Figure 2.1. a) Agglomerative clustering. b) Divisive clustering.

The methods for hierarchical clustering are divided into two subgroups: agglomerative and divisive methods. With agglomerative methods, units and clusters are progressively merged. Initially there are *n* clusters, representing the separate units and after the progressive merging, a single cluster is obtained, which incorporates all units. With divisive methods the approach is just the opposite – starting with one cluster, which incorporates all units, and after progressively partitioning it, *p* clusters are obtained, with each unit making up a separate cluster. The idea of clustering methods is illustrated in Figure 2.1.

The results of CA are usually represented using a dendrogram (tree diagram) which illustrates graphically the hierarchical structure generated by the similarity matrix and the clustering rules.

There are different strategies for merging objects in clusters and merging the clusters themselves. We will describe the most widely used linkage methods for hierarchical agglomerative clustering.

1) Within-Groups Linkage

The distance between any two clusters *A* and *B* is defined as the average of *nA.nB* distances between *nA* points from *A* and *nB* points from *B*:

$$D(A,B)=\frac{1}{n_A.n_B}\sum_{i=1}^{n_A}\sum_{j=1}^{n_B}d(x_i,x_j) \qquad (2.14)$$

where the sum includes all x_i from *A* and all x_j from *B*. Here $d(x_i,x_j)$ is the chosen distance between vectors x_i and x_j. At each step, the nearest two clusters are merged.

2) Between-Groups Linkage

This method is similar to within groups linkage, but it calculates all the possible distances between all points in the two clusters *A* and *B*, including the distances between the points within one and the same cluster. The formula is:

$$D(A,B)=\frac{1}{(n_A+n_B)(n_A+n_B-1)}\sum_{i,j}d(x_i,x_j), \qquad (2.15)$$

where the sum includes all points x_i and x_j from *A* and *B*, and $d(x_i, x_j)$ is the chosen distance. At each step, the nearest two clusters are merged.

3) Nearest Neighbor (Single Linkage)

With this method, the distance between two clusters *A* and *B* is equal to the minimum of all distances between the points in *A* and the points in *B*:

$$D(A,B) = \min\left\{d(x_i, x_j),\ x_i \in A,\ x_j \in B\right\} \tag{2.16}$$

4) Furthest Neighbor (Complete Linkage)

Here all distances are calculated using the formula:

$$D(A,B) = \max\left\{d(x_i, x_j),\ x_i \in A,\ x_j \in B\right\} \tag{2.17}$$

The two clusters with the smallest distance between them are merged into a new one.

5) Centroid Method

The mean vector $\overline{y}_A = \left(\sum_{i=1}^{n_A} y_i\right) / n_A$ is calculated for each cluster *A*, where n_A is the number of vectors (points) of *A*. Following which, all Euclidean distances between the mean vectors of each two clusters *A* and *B* are calculated

$$D(A,B) = (\overline{y}_A, \overline{y}_B) \tag{2.18}$$

The two clusters with the smallest distance between them are merged. The centroid of the new cluster *AB* is calculated using the formula

$$\overline{y}_{AB} = \frac{n_A \overline{y}_A + n_B \overline{y}_B}{n_A + n_B} \tag{2.19}$$

2.3.2.5. Determining the Number of Clusters

This is not a well-formalized procedure. The first approach is to read the dendrogram and to construct lines of intersection where the distance between one cluster and another is the greater, for example greater than at least 20% of the maximum scale.

Another ways is to compare the solution to another solution, obtained using a different clustering method and/or distance [8, 33, 34].

The procedure for determining the optimal number of clusters will be described in more detail further on, when CA is applied to the investigated data.

2.3.2.6. Cluster Analysis Results Validation

The following methods for results validation are recommended [8, 33, 34, 36]:

- Evaluation of the stability of the chosen cluster model by comparing the results obtained using several different methods
- Cross-validation, confirmed by randomly partitioning the sample into two non-complementary subsets, carrying out CA for each of them, and comparing the models
- Variable consistency check. When a variable (or an observation) is classified in different clusters with different methods, it should be excluded from the model as inconsistent or should be observed during subsequent statistical analyses.

2.3.2.7. Motivation for Using Cluster Analysis for Modeling MVLs

Applying CA to study laser sources, including lasers we are investigating will allow us to solve the following types of problems:

- Classifying input laser variables into macro categories and grouping them into several significant groups based on different similarity metrics
- Determining the positions of each independent variable and the groups within the data hierarchy
- Determining the degree of influence (distance) of the variables and the individual groups in relation to the dependent laser variables - laser efficiency and laser output power
- Using obtained classifications to construct models with the help of other statistical techniques, comparisons, and validations of models
- Solving individual problems for planning filtering and extremal experiments and predicting the experiment as a whole.

2.3.3. Factor Analysis

Factor analysis (FA) is a well-known statistical technique, which finds application in many fields, such as natural sciences, economics, psychology, sociology, etc.

We have to note that the most widely used type of FA is the so-called exploratory FA, which is applied in this study. The procedures for exploratory FA (determining the number of factors, factor extraction, rotation methods and more) are not sufficiently formalized and the obtained solution is not unique. The resulting factor model is considered to be satisfactory, if it can be interpreted sufficiently clearly in the context of the field it is applied to.

A large number of publications are devoted to FA, including the use of specialized software (for example, see [8, 33-37]).

2.3.3.1. General Characteristics

FA is a statistical technique designed to transform a correlated dataset into a new non-correlated dataset of artificial variables or factors, which describe as best as possible the variance of input data. Using this this method the initial number of variables is reduced by grouping those, which correlate in one factor, and singling out the non-correlating in different factors. Factors can be orthogonal or oblique.

Let $x_1, x_2, ..., x_p$ be the initial variables, each with n observations, constituting the data matrix $\mathbf{X} \in \mathbb{R}^{p \times n}$, in accordance with (2.7). Without loss of generality we will assume that $x_1, x_2, ..., x_p$ are standardized to z - scores and that their notation is the same. With FA, every x_j variable is modeled as a linear combination of k ($k < p$) number of n-dimensional factors, $F_1, F_2, ..., F_k$ plus errors $E_1, E_2, ..., E_k$ in the form

$$x_j = l_{j1}F_1 + l_{j2}F_2 + ... + l_{jk}F_k + E_j, \quad j = 1, 2, ..., p \qquad (2.20)$$

or $\mathbf{X} = \mathbf{LF} + \mathbf{E}$, where $\mathbf{F} \in \mathbb{R}^{k \times n}$ is the matrix of factor values, $\mathbf{L} \in \mathbb{R}^{p \times k}$ is the matrix of factor loadings, $\mathbf{E} \in \mathbb{R}^{p \times n}$ is matrix of the errors.

Mathematically from (2.20) each of the underlying factors $F_1, F_2, ..., F_k$ could be explicitly represented as a weighted sum of the p variables

$$F_i = w_{i1}x_1 + w_{i2}x_2 + ... + w_{ip}x_p, \quad i = 1, 2, ..., k \tag{2.21}$$

Or $\mathbf{F} = \mathbf{WX}$, where $\mathbf{W}$ is the weight matrix. Formula (2.21) shows the correlative relationship between $x_1, x_2, ..., x_p$ and how these are grouped in the factor F_i.

One of the main tasks in FA is to determine the number of factors k, which is strongly dependent on the degree of correlation between data.

FA allows the calculation of factor scores for each factor and the extraction of factor variables for each observation. They summarize the initial variables but unlike them, these cannot be observed and measured. Factor variables are represented in standard form by z – values, with a mean of zero and a standard deviation of one. These variables can be used in further statistical analyses instead of the initial variables.

A number of methods for extracting factors can be applied: principal component analysis (PCA), least squares method, maximum likelihood, principal axis factoring, etc. The most widely used of these is PCA. It has to be noted, that some researchers have contested its statistical qualities. For example, in one of the most frequently cited articles on the topic [28], through a detailed investigation of data from the field of psychology, using the Monte Carlo method, it is established that the maximum likelihood method is recommendable for relatively normally distributed data, and for severely violated data, for which there is an assumption of multivariate normality, it is recommendable to use the method of principal axis factors. We are tempted to add a comment that it seems the rule in force here is that the applicability of a given method should depend on the data, rather than the preference of the researcher.

2.3.3.2. Basic Assumptions for Applying Factor Analysis

Performing a factor analysis is based on the following assumptions [33, 34]:

- Random input data.
- Data must be of interval or ratio type. Category or nominal data are not suitable. Observations have to be independent.
- The recommended number of observations is at least 50 and the ratio $n / p > 10$ or at least $n / p > 5$ (see below).

- FA works with correlated variables. A variable, which does not correlate with the rest of the variables, needs to be excluded from FA (unique variable case). In subsequent analyses, such a variable could be used in conjunction with resulting factor variables.
- The data sample could be checked for factor analysis adequacy. This can be done for instance with the help of the Kaiser-Meyer-Olkin statistics (KMO test) which detect the absence of high multi-colinearity, requiring a score of >0.5, as well as using Bartlett's test in order to determine the sphericity of the data cloud.
- Multivariate normality is not considered as one of the critical assumptions of factor analysis.

2.3.3.3. Factor Analysis Using the Principal Component Analysis

We are going to summarize only the most widely used method for factor extraction which utilizes the principal component analysis (PCA) [33, 34].

Let us assume that all data are standardized using *z*-scores.

In order to apply PCA or FA, it is first necessary to find the correlation matrix

$$\mathbf{R} = \begin{pmatrix} 1 & r(x_1, x_2) & \dots & r(x_1, x_p) \\ r(x_2, x_1) & 1 & \dots & r(x_2, x_p) \\ \dots & \dots & \dots & \dots \\ r(x_p, x_1) & r(x_p, x_2) & \dots & 1 \end{pmatrix} \tag{2.22}$$

where $r(x_i, x_j)$ is the bivariate correlation coefficient of variables x_i, x_j.

The sum of the eigenvalues of this matrix is equal to the number of participating variables *p*. The relative eigenvalue λ_α shows the degree of participation of its corresponding factor F_α in the description of the general variance of the initial data.

If in the matrix (2.22), there are high correlation coefficients (except the ones along the principal diagonal), then perhaps our data are suitable for FA. One of the criteria for establishing this is the Bartlett's test of sphericity, to test the null hypothesis that the data sample was randomly chosen from a population in which the correlation matrix is almost an identity matrix (with all elements except the main diagonal are zeros).

If the correlation matrix (2.22) contains variables, which do not correlate with the others, these are removed from FA.

The PCA method transforms the data matrix **X** into another coordinate system by extracting exactly p new variables (called principal components), orthogonal to each other, thus distributing the entire variance of the data sample among them (analogically to (2.20)). This is displayed in Figure 2.2 for $p = 2$.

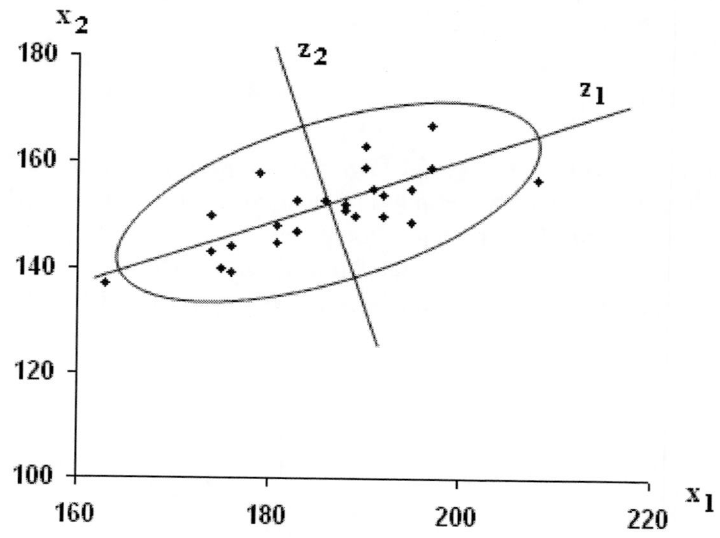

Figure 2.2. Example of the transformation of 2-coordinate system x_1, x_2 in a new z_1, z_2 orthogonal system using PCA method.

The first principal component is that which accounts for the largest amount of variance in the sample, the second principal component is that which accounts for the next largest amount of variance and is orthogonal to the first component and thus uncorrelated with it, and so on.

The next step is to determine how many and which of the obtained principal components should be chosen to conduct factor analysis. Usually the number of factors is chosen to be equal to the number of correlation matrix eigenvalues over 1.0, which is also known as Kaiser's rule [38]. The so-called 'scree plot' is also utilized, representing graphically the value of the eigenvalues and taking a number, which corresponds with the number of that eigenvalue for which the graph changes drastically.

However, specific examples were published, showing that due to the strong correlation between some of the factors and the dependent variable, some factors corresponding to eigenvalues much smaller than 1.0 could significantly influence the model [39]. What is more, the paper [40] presents examples, which show that the factor which corresponds to the smallest eigenvalue of the correlation matrix (2.22) has the strongest influence on the model, and the factors with the highest eigenvalues have no influence. This shows that a factor (even very small) which correlates with the dependent variable should not be dropped out of the analysis.

We are going to adhere to variance explained criteria, staying that a sufficient number of factors could be chosen so as to account for at least 90% of the overall variance of the data, with the goal of also including variables, which exhibit a strong correlation with the dependent variable. However, if the aim is to use the smallest possible number of factors to explain the variance, it is recommended to limit the selection even to 50% [33].

When the number of factors is known and fixed, they are extracted using Principal Component Analysis and the initial solution is obtained. Following that, the rotation is performed and the factors are found in the form of a "rotated solution." The factor rotation is performed in order to provide better differentiation of input variables into well distinctly different factors. Factors are usually rotated using the varimax method, but there are also other methods, such as quartimax, equamax, as well as oblique factor methods, for example oblimin, promax in SPSS, SAS, S-Plus etc.

The crucial point in FA is the analysis of the factor loadings matrix $\mathbf{L}^{rot} = (l_{ij}^{rot})$ of the rotated solution, called a rotation matrix. Factor loadings express the relationship between factors and input variables: they are the regression coefficients of input variables over the group of factors.

A variable, which has a factor loading with a big absolute value (for example over 0.5) can be grouped with the corresponding factor. When the absolute value of the loading is under the chosen limit, the participation of the variable is disregarded. A successful factor model requires a given variable to be grouped in this way only with a single factor.

2.3.3.4. Relationship between Sample Size and Factor Loading

There is no established rule as to what the size of the sample should be in order to perform FA. It is considered that each variable should have at least 10 or even 20 observations. For example, 8 variables require a sample size of at least 80.

Another important moment is the dependence between sample size and the value of factor loadings, which ensures that factor loadings are significant and variables with factor loadings over the limit are grouped together in the corresponding factor. Usually it is assumed that the significance limit is 0.5. More precise results are presented in [34, 41]. They are given in Table 2.4, at the significance level $\alpha = 0.05$. Consequently, for example, for a sample with a size of n = 100, only variables with factor loadings over 0.55 should be considered statistically significant for the corresponding factor.

Table 2.4. Recommended relation between sample size and factor loading in factor analysis

Sample size necessary for significance (at the 0.05 level)	Lower bound for a significant factor loading
350	0.30
250	0.35
200	0.40
150	0.50
100	0.55
85	0.60
70	0.65
60	0.70
50	0.75

2.3.3.5. Algorithm of Factor Analysis

- Normalization of all variables through transformation into *z*-scores
- Calculation of the correlation matrix and corresponding statistics
- Checking the FA adequacy
- Factor extraction using the Principal Component Analysis method or another technique for calculating communalities (accumulation of variances) and the distribution of total variance
- Choosing the number of factors
- Finding the initial factor solution
- Factor rotation and finding the rotation matrix
- Evaluation of the correctness of variable grouping in factors
- Calculation and storing of factor scores of factor variables for subsequent statistical analyses
- Interpreting the factors and results from FA.

2.3.3.6. Validation of Factor Analysis Results

Although FA is not a well-formalized technique, its individual stages, described above, need to satisfy a certain number of tests and requirements, which are used to validate the obtained factor model (see [5, 33, 34], etc.). These are the basic requirements:

- Used variables need to correlate with each other and with the dependent variables, i.e. bivariate correlation coefficients have to be relatively high. Otherwise FA is not advisable
- The statistical tests for adequacy of the FA for the given sample need to be performed: the score from the KMO test has to be >0.5, Bartlett's test of sphericity, also including the χ^2 test for multivariate normality needs to be statistically significant, i.e. Sig. < 0.05. The MSA (Measure of Sampling Adequacy) values for each variable can also be checked and their recommended value is over 0.5.
- Validation of the chosen number of factors using any method, for example by calculating the reproduced correlation residuals. To this end, the approximations for the variables of the calculated factors and factor loadings are found using formula (2.20), the correlation matrix is calculated, also known as a reproduced correlation matrix, and then it is compared with the initial correlation matrix. Residuals need to be sufficiently small, within 0.05.
- The grouping of factors in the rotation matrix needs to be correct. More exactly, each variable can only belong to one factor (to correlate strongly with it), and needs to have a weak correlation with the rest of the factors).

2.3.3.7. Motivation for Using Factor Analysis for Modeling MVLs

FA allows the construction of models based on experimental data for some MVLs. In the case of copper bromide lasers, FA is necessary because input variables are intercorrelated and the direct application of regression techniques is impossible or extremely difficult. FA makes it possible to solve the following types of problems:

- Classification of the independent variables of the laser system by grouping them into factors based on their mutual correlation

- Specifying the groups obtained using CA and excluding non-consistent input variables, with non-significant influence on the dependent variable
- Obtaining adequate multifactor models which describe a large percentage of the variance of the data
- Finding factor variables, which are independent from each other and suitable for subsequent statistical processing such as parametric or nonparametric regression analyses to estimate the dependencies.

2.3.4. Multiple Regression

Multiple regression analysis (RA) is one of the most powerful parametric techniques. It is used to develop models, describing explicitly the relationships between several independent variables $x_1, x_2, ..., x_p$ (called also predictors or explanatory variables) and one (or more) dependent on them variables (response). All the variables in a regression may be interval or ratio-scaled variables. Multiple regression model tries to explain the variance in the dependent variable on the variation in a set of relevant predictors thorough specific statistical tests. In this case we are looking for a general functional relationship

$$\hat{y} = \hat{f}(x_1, x_2, ..., x_p; a_1, a_2, ..., a_m) \qquad (2.4)$$

which expresses the influence of independent variables on the dependent variable. This equation is called a regression model or a regression equation of y with respect to $x_1, x_2, ..., x_p$, and $a_1, a_2, ..., a_m$ are the coefficients (parameters) of the regression, which are to be determined, as it was discussed in 2.1.3. We are going to construct linear and nonlinear regression equations of the type (2.4).

There are numerous publications dealing with RA. Practical applications (including using specialized software) are discussed in [5, 8, 33-36, 42-44].

2.3.4.1. Multiple Linear Regression

In multiple regression analysis (RA), the theoretical linear equation has the form

$$y = a_0 + a_1 x_1 + a_2 x_2 + ... + a_p x_p + e \tag{2.23}$$

or for all observations

$$y_i = a_0 + a_1 x_{i1} + a_2 x_{i2} + ... + a_p x_{ip} + e_i, \quad i = 1, 2, ..., n \tag{2.24}$$

with unknown regression coefficients $a = (a_0, a_1, ..., a_p)'$ and regression errors $e = (e_1, e_2, ..., e_n)'$. The coefficient a_j, $j = 1, 2, ..., p$ is called a partial regression coefficient and it characterizes the sensitivity of the variable y to the variance of x_j. The coefficient a_0 is called an intercept and it defines y, when all x_j variables are equal to zero.

Power terms can be added as independent variables in (2.23) to explore curvilinear forms. In addition, cross-product terms can be used to express interaction effects. In these cases, the equation is once again linear. For example, when $p = 2$, the model we are looking for assumes the following form

$$y = a_0 + a_1 x_1 + a_2 x_2 + a_3 x_1^2 + a_4 x_2^2 + a_5 x_1 x_2 + e.$$

By substituting $x_3 = x_1^2$, $x_4 = x_2^2$, $x_5 = x_1 x_2$ this equation becomes linear with respect to the new variables $x_1, x_2, ..., x_5$.

The regression parameters need to be evaluated, after having chosen a linear function of the model relationship (2.23). To this end, it is assumed that $n \geq p + 1$. If this is not satisfied and $n < p + 1$, then there are an infinite number of solutions to the system (2.24) for the data $(x_{1j}, x_{2j}, ..., x_{np}, y_i)$. If $n = p + 1$, the system (2.24) has a single solution for a determinant, different from zero.

It is impossible to obtain the real values of the regression coefficients for a given sample. Instead of the theoretical equation (2.23), the empirical regression equation is evaluated, which is represented as

$$y = b_0 + b_1 x_1 + b_2 x_2 + ... + b_p x_p + \varepsilon \tag{2.25}$$

or for each case we have

$$y_i = b_0 + b_1 x_{i1} + b_2 x_{i2} + ... + b_p x_{ip} + \varepsilon_i, \quad i = 1, 2, ..., n \tag{2.26}$$

with empirical regression coefficients $b = (b_0, b_1, ..., b_p)'$ and errors $\varepsilon = (\varepsilon_1, \varepsilon_2, ..., \varepsilon_n)'$.

In order to estimate the coefficients of multiple linear regression a large number of methods can be used. We are going to describe in more detail only the ordinary least squares method (OLS). As a matter of fact, with some datasets, it is possible to also try and use ridge regression, partial least squares or some other alternative to least squares [8, 36, 45-47].

Let us assume that the conditions for applying the OLS method have been met. In this case, empirical regression coefficients $b = (b_0, b_1, ..., b_p)'$ are unbiased, effective and consistent estimates of the theoretical parameters $a = (a_0, a_1, ..., a_p)'$.

In order to compare the relative effect of individual quantities they are transformed into *z*-values which allows the formulation of the standardized regression equation

$$\frac{y - \overline{y}}{s_y} = \beta_1 \frac{x_1 - \overline{x}_1}{s_1} + \beta_2 \frac{x_2 - \overline{x}_2}{s_2} + ... + \beta_p \frac{x_p - \overline{x}_p}{s_p} + \varepsilon \tag{2.27}$$

or

$$y = \beta_1 z_1 + \beta_2 z_2 + ... + \beta_p z_p + \varepsilon \tag{2.28}$$

Here, the coefficients $\beta_1, \beta_2, ..., \beta_p$ are called standardized regression coefficients, other notations are given in 2.3.1.

After obtaining the regression coefficients $b_0, b_1, ..., b_p$ or $\beta_1, \beta_2, ..., \beta_p$ they are substituted in the equation and used to evaluate and predict the dependent variable for given values of the independent variables

$$\hat{y}_i = b_0 + b_1 x_{i1} + b_2 x_{i2} + ... + b_p x_{ip}, \quad i = 1, 2, ..., n \tag{2.29}$$

or in standardized form:

$$\hat{y} = \beta_1 z_1 + \beta_2 z_2 + ... + \beta_p z_p \tag{2.30}$$

We denote the residuals in the model (2.29) with

$$\varepsilon_i = y_i - \hat{y}_i = y_i - b_0 - b_1 x_{i1} - b_2 x_{i2} - \ldots - b_p x_{ip}, \qquad i = 1, 2, \ldots, n$$

or

$$\varepsilon = y - \hat{y}$$

By using notations (2.7), (2.8) it is immediately becomes apparent that

$$\varepsilon = y - \tilde{\mathbf{X}}b$$

OLS is used to find the set of regression coefficients $b = (b_0, b_1, \ldots, b_p)'$, which minimizes the residual sum of squares

$$\Phi(b_0, b_1, \ldots, b_p) = \sum_{i=1}^{n} \varepsilon_i^2 = \sum_{i=1}^{n} \left(y_i - b_0 - b_1 x_{i1} - b_2 x_{i2} - \ldots - b_p x_{ip} \right)^2 \quad (2.31)$$

In matrix form, the function (2.31) is represented as follows

$$\Phi = \varepsilon'\varepsilon = \left(y - \tilde{\mathbf{X}}b \right)' \left(y - \tilde{\mathbf{X}}b \right) = y'y - b'\tilde{\mathbf{X}}'y - y'\tilde{\mathbf{X}}b + b'\tilde{\mathbf{X}}'\tilde{\mathbf{X}}b =$$
$$y'y - 2b'\tilde{\mathbf{X}}'y + b'\tilde{\mathbf{X}}'\tilde{\mathbf{X}}b$$

The necessary condition for the existence of an extreme value of the function Φ is that all partial derivatives with respect to b_j need to be equal to zero, i.e. $\frac{\partial \Phi}{\partial b_j} = 0$. It is not difficult to show that

$$\frac{\partial \Phi}{\partial b_j} = -2\tilde{\mathbf{X}}'y + 2(\tilde{\mathbf{X}}'\tilde{\mathbf{X}})b, \quad j = 0, 1, \ldots, p$$

In this way, the problem is reduced to solving the algebraic system

$$(\tilde{\mathbf{X}}'\tilde{\mathbf{X}})b = \tilde{\mathbf{X}}'y \quad (2.32)$$

If the determinant of the matrix $\tilde{\mathbf{X}}'\tilde{\mathbf{X}}$ is significantly different from zero (the matrix is nonsingular), then from (2.32) we can express the sought empirical coefficients of the multiple linear regression in matrix form

$$b = (\tilde{\mathbf{X}}'\tilde{\mathbf{X}})^{-1}\tilde{\mathbf{X}}'y \qquad (2.33)$$

where $(\tilde{\mathbf{X}}'\tilde{\mathbf{X}})^{-1}$ is the inverse matrix of $\tilde{\mathbf{X}}'\tilde{\mathbf{X}}$. The resulting vector (2.33) gives the minimum of $p=3$ (2.31) and its elements are called the least squares estimates of the theoretical regression coefficients. It is important to realize that the least squares estimates are only appropriate when the model (2.25) and the assumptions on OLS are valid regarding the Gauss-Markov Theorem.

In practice, usually the system (2.32) is solved directly by some numerical method, without inverting the matrix $\tilde{\mathbf{X}}'\tilde{\mathbf{X}}$.

For example, for $p=3$, in order to determine the regression coefficients $b=(b_0,b_1,b_2,b_3)'$ the system (2.32) is:

$$\begin{aligned}
&b_0 n + b_1\sum_{i=1}^{n} x_{i1} + b_2\sum_{i=1}^{n} x_{i2} + b_3\sum_{i=1}^{n} x_{i3} = \sum_{i=1}^{n} y_i\\
&b_0\sum_{i=1}^{n} x_{i1} + b_1\sum_{i=1}^{n} x_{i2}^2 + b_2\sum_{i=1}^{n} x_{i1}x_{i2} + b_3\sum_{i=1}^{n} x_{i1}x_{i3} = \sum_{i=1}^{n} x_{i1}y_i\\
&b_0\sum_{i=1}^{n} x_{i2} + b_1\sum_{i=1}^{n} x_{i1}x_{i2} + b_2\sum_{i=1}^{n} x_{i2}^2 + b_3\sum_{i=1}^{n} x_{i2}x_{i3} = \sum_{i=1}^{n} x_{i2}y_i\\
&b_0\sum_{i=1}^{n} x_{i3} + b_1\sum_{i=1}^{n} x_{i1}x_{i3} + b_2\sum_{i=1}^{n} x_{i2}x_{i3} + b_3\sum_{i=1}^{n} x_{i3}^2 = \sum_{i=1}^{n} x_{i3}y_i
\end{aligned}$$

The variance and standard errors of the regression coefficients need to be known in order to analyze the accuracy of the various statistical characteristics and to check hypotheses.

It is well known, that the variance of errors is presented by the formula

$$D(e_i) = \sigma^2 z'_{jj}$$

where z'_{jj} is the j-th diagonal element of the matrix $Z^{-1} = (\tilde{\mathbf{X}}'\tilde{\mathbf{X}})^{-1}$. Since the exact value of the variance σ^2 is unknown, it is substituted with its unbiased estimate

$$S^2 = \frac{\sum_{i=1}^{n} \varepsilon_i^2}{n-p-1}$$

where p is the number of predictor variables and $n-p-1$ is the degree of freedom. The standard variance error is $S = \sqrt{S^2}$.

The following equations are valid for the sample variance of the empirical coefficients of the regression for a given sample:

$$S_{b_j}^2 = S^2 z'_{jj} = \frac{\sum_{i=1}^{n} \varepsilon_i^2}{n-p-1} z'_{jj}, \quad j = 0,1,2,...,p$$

The standard error of the regression coefficient b_j is $S_{b_j} = \sqrt{S_{b_j}^2}$.

A detailed mathematical description of the statistical tests, the confidence and prediction intervals, as well as other statistical indices of multiple regression are described in numerous sources, including software, for example, see [5,8,33-36].

2.3.4.2. Nonlinear Regression

Parametric nonlinear regression can be used when the data do not indicate a linear relationship and conducting multiple linear regression does not produce good results. To be more exact, in this case it is assumed that the model (2.4) is nonlinear with respect to the regression coefficients. The estimates of the model parameters are chosen to minimize the χ^2 merit function given by the sum of squared residuals $\sum_i \varepsilon_i^2$. They can be determined using various optimization methods, such as the least-squares method, weighted least-squares, gradient method (steepest descent), Newton method, sequential quadratic programming, etc.

As with the parametric linear regression, the type of nonlinear dependence needs to be specified beforehand. For example

$$y = \beta_0 + \beta_1 x^2 + \beta_2 \exp(\beta_3 + \beta_4 x) + \varepsilon$$

where $\beta_0, \beta_1, \beta_2, \beta_3, \beta_4$ are the unknown parameters. Other problem is to find a feasible starting points for numerical solving of the respective optimization problem. However, some software packages as Wolfram *Mathematica* solve automatically this problem (see build-in function NonlinearModelFit, [48]).

In many cases, the type of the nonlinear expression is determined by the behavior of the data, for example, when they are studied using graphs. For a more in-depth understanding of nonlinear models, both theoretically and with examples, see [8, 30, 36, 42].

2.3.4.3. Basic Requirements for Applying Regression Analysis

We are going to describe only the basic rules for using RA.

- Input data have to be random interval or ratio-scaled data. Categorical data need to be transformed into binary in advance. Observations must be independent.
- For a given independent variable, the dependent variable needs to have a normal distribution or an approximately normal one. The requirements for multiple RA are analogical.
- With linear regression, it is assumed that the general dependence is linear or approximately linear.
- In nonlinear regression, it is assumed that the selected functions describe adequately the behavior of the dataset.

In the classic case, applying RA necessitates very strict requirements for the data, which are difficult to satisfy in practice. For multiple linear regression the distributions should be at least close to normal, and that the validity of obtained models needs to be examined using graphics, sample modification, comparisons to results from other methods, and more (see [33, 34, 42, 43]).

2.3.4.4. Validation of Regression Analysis Results

The first important key step in validating the regression model is the check of the theoretical assumptions. The main ones are (see also examples in [33, 34]):

- Presence of causality. Variables are to be classified as dependent and independent.
- Investigation of variable consistency and if possible inclusion of all relevant variables in the model, so as not to omit the influence of any other variable, which is not accounted for by the analysis. Such a problem is usually resolved by a stepwise regression.
- Check for a linear or close to linear relationship between independent variables and the dependent variable in the case of multiple linear regression.
- Investigation for the presence of additional dependencies between the independent variables and the dependent variable.
- Adequate sample size n, with the standard minimum requirement being that $n > 10\,p$, where p is the number of predictors.

The second key is the validation of numerous statistical tests for regression models. When they are successful, the regression model is considered statistically significant and suitable for practical application and predictions. This includes the check of the statistical significance of the regression equation coefficients and the check of the goodness of fit of the model as a general attribute of the regression equation.

The significance of each coefficient is checked along with the null hypothesis that the respective population coefficient is zero. Coefficients with only a slight statistical significance are excluded from the analysis since this would not lead to any noticeable loss of information.

When analyzing the overall adequacy of the model, the most important quantity is the coefficient of determination R-square, considered as a goodness of fit measure. It is used in the context of statistical models whose main purpose is the prediction of future outcomes on the basis of the available dataset. Its value provides a measure of how well future outcomes are likely to be predicted by the model. In addition, it is also important to test the null hypothesis that the obtained regression equation does not describe the variability of the dependent variable.

Numerically, the coefficient of determination R-square is the square of the multiple correlation coefficient R and is calculated by the expression

$$R^2 = 1 - \frac{\sum_{i=1}^{n} \varepsilon_i^2}{\sum_{i=1}^{n} (\hat{y}_i - \overline{y})^2}, \quad 0 \le R^2 \le 1$$

where $\overline{y}$ is the mean value of y_i.

The coefficient of determination measures the proportion of the variability in the dependent variable that is explained by variations in the predictors of the model. The closer R-square is to 1, the better the model. Of course, in this case the possibility of overestimation, i.e. unnecessarily complicating the model in order to achieve a higher value for the coefficient of determination, also has to be taken into account.

It is also recommended to calculate the so called adjusted R-square coefficient. It takes into account the number p of relevant independent variables in a multiple regression. Its value may be found by the expression

$$R_{adj}^2 = 1 - \frac{\sum_{i=1}^{n} \varepsilon_i^2 /(n-p-1)}{\sum_{i=1}^{n} (\hat{y}_i - \overline{y}_i)^2 /(n-1)}, \quad R_{adj}^2 < R^2$$

The third key is the diagnostics of the regression residuals. This is an indispensable part of the model validation test. The regression residuals have to satisfy certain statistical properties. Only if these conditions are satisfied, we can be confident in our model as well as in the interpretation of the results. The mentioned conditions are based upon the complex mathematical theory and can be found in almost every statistics textbook.

Each residual is the difference between the observed value of the dependent variable and its corresponding predicted value. Residuals are that which cannot be explained by the regression equation and they are classified as "noise" or errors, when the regression equation has been found to be correct.

When conducting RA, it is assumed that the residuals $\varepsilon_i = y_i - \hat{y}_i$ are independent, have zero means, the same constant variance and are normally distributed. The confirmation of these assumptions is used as a proof that the model has been constructed correctly.

To this end, the following four characteristics are checked:

1) Independence. This means that each observation is independent of the others.
2) Zero means. For example, shown using a histogram of the residuals.
3) Normality. This assumption may be tested by the histogram of regression standardized residuals, the normal probability plot (Normal P-P plot) or it could be checked using formal Kolmogorov-Smirnov or Shapiro-Wilk tests. If the general residuals graph is considered, it is immediately obvious which observations are sharply separated (outside from the general behavior pattern). It should be analyzed what the reason for this is and if necessary a new model should be constructed, excluding such observations.
4) Homoscedasticity. This means that the residual has the same variance for every value of the independent variable. Otherwise, the model displays heteroscedasticity. This assumption can be verified by constructing the plot of Regression Standardized Predicted Values (ZPRED in SPSS) versus Regression Standardized Residuals (ZREZID). There should be no discernible patterns. The patterns of triangles or diamonds indicate, that condition 4) is not satisfied. The reason for this could be the incorrect form of the regression model, the lack of reliable variables or the use a mis-measured independent variable. It is also recommended to construct the plots of each independent and dependent variable versus the Standardized Residuals. They should not demonstrate dependences. The presence or absence of patters would indicate respectively heteroscedasticity or homoscedasticity within the model.

In summary, we are going to list only the main statistical tests and indices for validity of RA (at the 0.05 significance level):

- Significance of the model (table 'ANOVA'). The level of significance of the model as a whole, evaluated by an F statistic needs to be less than 0.05. Typically, if 'Sig." is greater than 0.05, the conclusion is that the model could not fit the data
- Fitting the data (Model summary). The coefficient of determination R-square and the absolute value of multidimensional correlation coefficient R have to be close to 1 and preferably over 0.5. Standard Error of the Estimate has to be not very big relatively to the mean of the predicted values of the dependent variable.

- Significance of the regression coefficients (Table 'Coefficients'). The level of significance of any of regression coefficients in the model needs to be less than 0.05 (*Sig.* <0.05). If some significance value is more than 0.5, then the coefficient estimate is not reliable and could drop from the model.
- Diagnostics of the residuals. The results of the regression model are valid, if the residuals are independent, have zero mean, are normally distributed and satisfy the homoscedasticity property.

2.3.4.5. Motivation for Using Regression Analysis for Modeling MVLs

With the help of the multiple RA, we can obtain the explicit form of the dependence of output variables on independent input laser variables and/or on grouping factors. In particular, RA assists in solving the following problems:

- Obtaining parametric linear and quasilinear models, describing the dependencies between input and output laser variables.
- Using the obtained models to calculate the values of the dependent variables and to predict the experiment.
- Using the models as a reference point for comparing results from nonparametric regression models.

2.3.5. Principal Component Regression

In statistics, there are different modifications of multiple regression analysis, the well-known being: Principal Component Regression, Partial Least-Squares Regression, and Ridge Regression [8, 45-47].

In addition to classical multiple regression, this study makes use of Principal Component Regression (PCR) firstly proposed in [46]. This method is used when explanatory variables are multicollinear. In this case, the matrix $\tilde{\mathbf{X}}'\tilde{\mathbf{X}}$ in (2.32) used in the ordinary least squares is singular. The inverse matrix $(\tilde{\mathbf{X}}'\tilde{\mathbf{X}})^{-1}$ is also utilized when calculating the standard error of the regression coefficients and of their correlation matrix. If there is a singularity, the results would exhibit big calculation errors.

In order to avoid this problem, with the PCR method, the initial independent variables $x_j,\ j=1,2,...,p$ are transformed into a set of non-correlating variables, which are their linear combinations. This is achieved through factor analysis with extraction of factors via Principal Component

Analysis. In fact, PCR consists of two steps. In the first step, one applies PCA on the input variables and finds uncorrelated orthogonal each to other principal components. In the second step, one chooses a selection of principal components to regress the dependent variable by OLS. What is important is to retain appropriate principal components providing the maximum variation of **X** which optimizes the predictive ability of the model. In principle, the PCR estimates of the obtained regression coefficients are biased and the bias is smaller when the involved factors account as big as possible variation in the data [47]. In some improved version of PCR the corresponding estimates of the regression parameters are conditionally unbiased [49].

2.4. Nonparametric Regression by Using Multivariate Adaptive Regression Splines (MARS)

During the past ten years, nonparametric regression has become increasingly popular, but the field is still underdeveloped compared to classical regression methods. Nonparametric regression analysis traces the dependence of a response variable on one or several predictors without specifying in advance the function that relates the predictors to the response. This way, the user essentially lets the data "model themselves". The most popular nonparametric regression methods include different types of smoothing techniques: LOWESS (Local WEighted Scatterplot Smoother) which stands for "locally weighted regression", kernel estimation, local polynomial regression, smoothing splines, additive nonparametric regression, etc. (see below) for studying data and patterns in data, without making an assumption about the type of distribution [8, 9, 36].

One of the most advanced local nonparametric methods, namely MARS, has been used in this study.

2.4.1. General Characteristics of MARS

The method of MARS is a kind of nonparametric regression technique for studying relationships. It was initially developed for data mining with large multidimensional datasets, but it displayed excellent qualities in a number of other fields. Today it is successfully used as a prediction and description technique in economics, sociology, ecology, geographical information systems, meteorology, engineering etc.

Being a nonparametric regression method, MARS does not impose strong limitations on the distribution of data, characteristic of classic parametric methods such as multiple regression or logistic regression. It is comparable to other nonparametric methods used for describing complex relationships between variables, namely Classification and Regression Trees - CART, Artificial Neural Networks (ANN), additive models, in particular the generalized additive model (GAM) etc. MARS has a number of advantages over these methods.

In this study, MARS was chosen over other nonparametric methods for modeling data for the following main reasons [11, 50]:

- MARS is suitable for processing not only big but also small data arrays, which are often found in engineering problems, when compared to CART, ANN and other techniques; the sample size limit is $n > 50$.
- The constructed model can be used with new data in order to predict the approximated function, including data outside the given intervals for the variables.
- It works with continuous data type.
- Easier interpretation of the results and assessment of the influence of individual variables on the dependent variable.
- It is possible to directly compare the results, obtained by RA.
- Much higher data processing speed (for example, between 100 – 1000 times faster than ANN).

2.4.2. Brief Description of the Method

The mathematical basis of MARS was developed by the American statistician Jerome Friedman in 1990-1991 [10]. The algorithms created by Friedman and their first working program versions have been integrated in the currently existing MARS software from SalfordSysrems. This product has gained popularity and has been applied with increasing success in the last few years [11, 50]. Other software packages providing MARS technique are as follows: STATISTICA Data Miner from StatSoft, ARESLab toolbox for Matlab and several R functions (polymars, mars, earth).

In essence, MARS generates adaptive models through partially linear regressions, i.e. data nonlinearities are approximated using separate sloped intercepts in different subintervals of the set defined for each predictor variable. A broken line is used, instead of looking for one common regression curve fitting the data. The slope of the regression line varies from one interval to another at the so-called nodes.

The knot is a basic element of the model. It shows where the behavior of the function changes. In the classic spline, nodes are given in advance, while with MARS, the most suitable place for them is determined using a fast algorithm when certain suitable optimization conditions are met. The initial node of the searching procedure is always at the lowest value of the predictor (the minimum). In addition, it is possible to find new relationships between variables and to determine new variables to be included in the model. In practice, determining the best distribution of the nodes for a large number of variables, especially as they are usually an unknown number, is a very complex task, which requires intensive calculations. This problem is solved by using the so-called "forward-backward stepwise procedure" and intermediate (fictitious) points for every distinct predictor variable.

The other basic element of MARS is the basis function for transformation of predictors. Before the processing of the data, by default, all variables are standardized into *z*-scores.

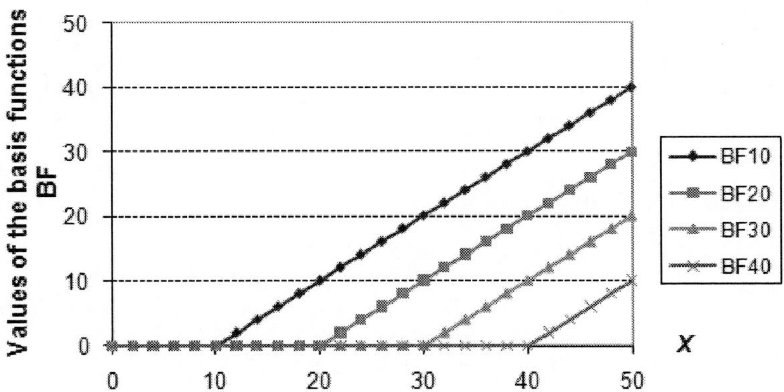

Figure 2.3. Graphics of four basis "hockey stick" functions for the predicator variable X, defined in the interval [0, 50].

The basis function is called a "hockey stick" or "hinge function" and has one of the following form:

$$BF(X) = \max(0, X - c)$$
$$BF(X) = \max(0, c - X) \qquad (2.34)$$
$$BF(X) = 1,$$

where c is a knot, with which the function X is mapped in X^*. An example with four basis functions is given in Figure 2.3. In fact, such a function is generated for each value of X.

In the case of the function $BF20$ in Fig. 2.3 we have the following transformation for X:

$$BF20(X) = \max(0, X - 20)$$

If in the regression equation X is replaced with the basis function $BF20$ we will get

$$y = const + b_1 BF20(X) + error$$

The spline will be written in the following form:

$$\hat{y} = \begin{cases} const & X \in [0, 20] \\ const + b_1(X - 20) & X \in [20, 50] \end{cases}$$

If a second function is also used, for example $BF30(X) = \max(0, X - 30)$, the regression function will assume the form

$$y = const + b_1 BF20 + b_2 BF30 + error$$

For $X \geq 30$ we get

$$\hat{y} = const + b_1(X - 20) + b_2(X - 30) = const + 10b_1 + (b_1 + b_2)(X - 30)$$

For convenience, the spline is written as

$$\hat{y} = \begin{cases} const & X \in [0,20] \\ const + b_1(X-20) & X \in [20,30] \\ const_1 + (b_1 + b_2)(X-30) & X \in [30,50] \end{cases}$$

with slopes 0, b_1 and $(b_1 + b_2)$. Here, the nodes are points 0, 20 and 30. A more complex linear spline can be constructed analogically. The coefficients of the spline are determined by additional optimization conditions, minimizing the total regression error when using different algorithms, adaptively in relation to the data. An example with linear MARS model with two knots is shown in Figure 2.4.

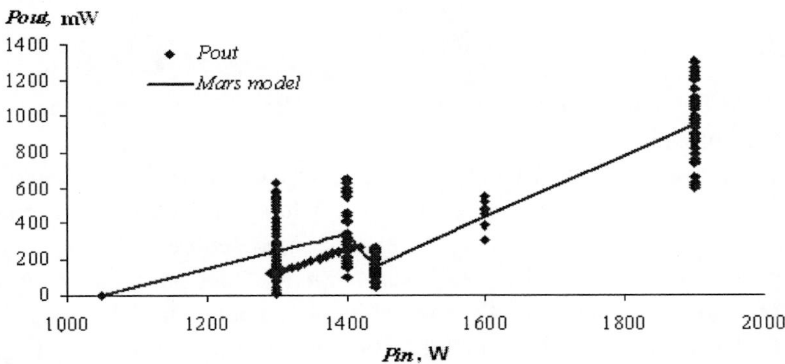

Figure 2.4. Partially linear regression curve with two nodes (continuous line) and observed data (points).

The general MARS model has the form

$$\hat{y} = \mu(X) = b_0 + \sum_{j=1}^{s} b_j BF_j(X) \tag{2.35}$$

where $BF_j(x)$ are the basis functions, b_j are the constant coefficients, s is the number of basic functions.

When we want to build a model, which also includes higher order interactions, the corresponding basis functions in (2.35) are the products of linear hinge functions. For example, a second order basis function is

$$BF_k(X) = BF_j(X).BF_i(X) \tag{2.36}$$

Another important feature of MARS is the application of different smoothing methods. To name a few: smoothing means, smoothing medians, weighted least squares method, LOWESS, etc.

We are also going to briefly look at some techniques for testing, validation and avoidance of an overfitting of the model. First, the number of candidate basis functions, for which it is assumed to be part of the "best model", is estimated. After that, their number is at least doubled. Basis functions are consecutively introduced into the model (2.35) in a forwards-stepwise procedure. At each step an input variable in the form (2.34) or (2.36) and a knot are chosen by minimizing the residual sum of squares

$$\sum_{i=1}^{n} \left(y_i - \mu_m(x_i)\right)^2, \qquad m = 1, 2, \ldots, M$$

where m is the number of the current model and M is the maximum number of "forward" models.

Then a backwards-stepwise procedure is applied to prevent the overfitting of the model. Now the goal is to exclude those functions, which contribute the least to the accuracy of the model. The exclusion procedure is iterative and can be described as follows. In the model with the biggest number of basis functions, MARS determines the one which contributes the least to the least squares criterion after which that function is removed. The new model is calculated and in the same way the next least influential basis function is removed and so on.

We have to specify that the "naïve model" suggested by MARS has the biggest coefficient of determination R^2 and corresponds to the model with the maximum number of functions. At backwards procedure MARS tries to find "the best" model which minimizes the General Cross-Validation criterion (GCV). The GCV criterion was introduced by one of the pioneers of splines, Grace Wahba [51] and extended by J. H. Friedman [10] for MARS. It has the form (see also [8, 36, 50]):

$$GCV(m) = \frac{1}{n} \sum_{i=1}^{n} \left(y_i - \hat{\mu}_m(x_i)\right)^2 / \left(1 - \frac{C(m)}{n}\right)^2, \quad m = 1, 2, \ldots, M \qquad (2.37)$$

where $\hat{\mu}_m(x)$ is the fitted value of the current model $\mu_m(x)$ with s terms and $C(m)$ is a complexity cost function [51], defined in the MARS context as

$$C(m) = s + \delta(s-1)/2, \quad 2 \le \delta \le 3 .$$

Note that $(s-1)/2$ is the number of hinge-function knots, so the formula penalizes the addition of knots.

2.4.3. Motivation for Using MARS in Modeling MVLs

The application of a nonparametric method for modeling the characteristics of metal vapor lasers is motivated by several reasons:

- For some data, the conditions for applying traditional methods cannot be met, for example, the data on the ultraviolet copper bromide vapor laser.
- In many cases, data show local variations and are difficult to describe using a general parameter curve.
- When MARS is used, it is possible to perform comparisons with models, obtained using parametric models.
- It is very easy to interpret model results.
- Predictions are easily made in accordance with a given local variance region of the predictors.

REFERENCES

[1] D. C. Montgomery and G. C. Runger, *Applied statistics and probability for engineers*, 3rd ed., John Wiley and Sons, New York, 2003.

[2] R. A. Fisher, *The Design of experiments*, Oliver and Boyd, Edinburgh, 1935.

[3] NIST/SEMATECH *e-Handbook of Statistical Methods*, http://www.itl.nist.gov/div898/handbook/.

[4] D. C. Montgomery, *Design and analysis of experiments*, 5th ed., Wiley, New York, 2001.

[5] T. P. Ryan, *Modern engineering statistics*, John Wiley and Sons, Inc., Hoboken, New Jersey, 2007.

[6] M. Ross, *Introduction to probability and statistics for engineers and scientists*, 3rd ed., Elsevier Academic Press, Amsterdam, New York, 2004.

[7] L. Breiman, J. H. Friedman, R. A. Olshen and C. J. Stone, *Classification and regression trees*, Wadsworth and Brooks/Cole, Monterey, CA, 1984.

[8] T. Hastie, R. Tibshirani, and J. Friedman, *The Elements of Statistical Learning: Data Mining, Inference and Prediction*, 2nd ed., Springer, New York, 2009.

[9] T. J. Hastie and R. J. Tibshirani*, Generalized additive models*, Chapman and Hall/CRC, 1990.

[10] J. H. Friedman, *Multivariate adaptive regression splines (with discussion), The Annals of Statistics*, 19(1) (1991) 1-141.

[11] D. Steinberg, B. Bernstein, P. Colla and K. Martin, *MARS User Guide, Salford Systems*, San Diego, CA, 2001.

[12] D.N. Astadjov, N.K.Vuchkov and N.V. Sabotinov, *Parametric study of the CuBr laser with hidrogen additives, IEEE J. of Quant. Electr.* (9) 24 (1988) 1926-1935.

[13] D. N. Astadjov, N. V. Sabotinov and N. K. Vuchkov, *Effect of hydrogen on CuBr laser power and efficiency, Opt. Commun.* (4) 56 (1985) 279–282.

[14] D. N. Astadjov, K. D. Dimitrov, C. E. Little and N. V. Sabotinov, *A CuBr laser with 1.4 W/cm^3 average output power, IEEE J. Quantum Electron.*, (6) 30 (1994) 1358–1360.

[15] N. K. Vuchkov, D. N. Astadjov and N. V. Sabotinov, *A new circuit for CuBr laser excitation, Opt. Quantum Electron*. 23 (1991) S549–S553.

[16] N. V. Sabotinov, *Copper bromide lasers*. In: Pulsed metal vapour lasers, Eds. C. E. Little and N. V. Sabotinov, NATO ASI series, Disarmament Technologies-5, Kluwer Academic Publishers, Dordrecht, 996, 113–124.

[17] K.D. Dimitrov and N.V. Sabotinov, *High-power and high-efficiency copper bromide vapor laser*, 3052 *SPIE* (1996) 126–130.

[18] D. N. Astadjov, K. D. Dimitrov, D. R. Jones, V. L. Kirkov, L. Little, C. E. Little, N. V. Sabotinov, and N. K. Vuchkov, *Influence on operating characteristics of scaling sealed-off CuBr lasers in active length, Opt. Commun.* 135 (1997) 289–294.

[19] D. N. Astadjov, K. D. Dimitrov, D. R. Jones, V. K. Kirkov, C. E. Little, N. V. Sabotinov, et al., *Copper bromide laser of 120-W average output power, IEEE J. Quantum Electron*., (5) 33 (1997) 705–709.

[20] V. M. Stoilov, D. N. Astadjov, N. K. Vuchkov and N. V. Sabotinov, *High spatial intensity 10 W–CuBr laser with hydrogen additives, Opt. Quantum Electron.*, 32 (2000) 1209–1217.

[21] *NATO contract SfP*, 97 2685 (50W copper bromide laser), 2000.

[22] N. P. Denev, D. N. Astadjov and N. V. Sabotinov, *Analysis of the copper bromide laser efficiency*, In: Proc. of Fourth International Symposium on Laser Technologies and Lasers '2005, Plovdiv, Bulgaria, 2006, 153–156.

[23] N. K. Vuchkov, K. A. Temelkov and N. V. Sabotinov, *UV Lasing on Cu+ in a Ne-CuBr Pulsed Longitudinal Discharge, IEEE J. Quantum Electron.* 35(12) (1999) 1799-1804.

[24] N. K. Vuchkov, K. A. Temelkov, P. V. Zahariev and N. V. Sabotinov, *Influence of the active zone diameter on the UV-ion Ne-CuBr laser performance, IEEE J. Quantum Electron.*, 37(12) (2001) 1538-1546.

[25] N. K. Vuchkov, *UV copper ion laser in Ne-CuBr pulse-longitudinal discharge*, In Advances in Laser and Optics Research, Ed. W.T. Arkin, Nova Science Publishers Inc., New York, 2002, 1-33.

[26] N. K. Vuchkov, K. A. Temelkov and N. V. Sabotinov, *Effect of hydrogen on the average output of the UV Cu+ Ne-CuBr laser, IEEE J. Quantum Electron.*, 41(1) (2005) 62-65.

[27] N. Vuchkov, *High discharge tube resource of the UV Cu+ Ne-CuBr laser and some applications*, In: *New development in lasers and electric-optics research*, ed. W. T. Arkin, Nova Science Publishers, New York, 2006, 41-74.

[28] L. R. Fabrigar, D. T. Wegener, R. C. MacCallum and E. J. Strahan, *Evaluating the use of exploratory factor analysis in psychological research*, *Psychological Methods*, 4(3) (1999) 272-299.

[29] B. Efron and R.J. Tibshirani, *An introduction to the bootstrap*, Chapman and Hall, London, 1993.

[30] D. M. Bates and D. G. Watts, *Nonlinear regression analysis and its applications*, Willey, New York, 1988.

[31] G. E. P. Box and D. R. Cox, *An analysis of transformations*, *Journal of the Royal Statistical Society,* Ser. B, 26 (1964) 211-252.

[32] K. Yeo and R. A. Johnson, *A new family of power transformations to improve normality or symmetry*, Biometrika, Oxford Press, 87(4) (2000) 954–959.

[33] C. Rencher, *Methods of multivariate analysis*, 2 ed., John Wiley, New York, 2002.

[34] W. Janssens, K. Wijnen, P. De Pelsmacker, P. Van Kenhove, *Marketing research with SPSS*, Prentice Hall, Pearson Education Ltd, Harlow, 2008.

[35] http://www.spss.com/software/statistics/stats-pro/, IBM SPSS Statistics.

[36] J. Izenman, *Modern multivariate statistical techniques: regression, classification, and manifold learning*, Springer, New York, 2008.

[37] J. Kim and Ch. W. Mueller, *Factor analysis: Statistical methods and practical issue*, Eleventh Printing, Sage Publication, 1986.

[38] H. F. Kaiser, *The varimax criterion for analytic rotation in factor analysis*, *Psychometrika,* 23 (1958) 187-200.

[39] T. Jolliffe, A note on the use of Principal Components in regression, *Journal of the Royal Statistical Society*, Series C (Applied Statistics), 31(3) (1982) 300–303.

[40] S. Hadi and R. F. Ling, *Some cautionary notes on the use of Principal Components Regression, The American Statistician,* 52 (1998) 15-19.

[41] BMDP Statistical Software Inc., *Computation with solo power analysi,.* LA: BMDP Statistical Software, 1993.

[42] N. Draper and H. Smith, *Applied Regression Analysis*, Wiley, New York, 1981.

[43] S. Weisberg, *Applied linear regression*, 3rd ed., Wiley-Interscience, Hoboken, New Jersey, 2005.

[44] A. Afifi and Y. Clark, *Computer-aided multivariate analysis*, Lifetime Learning Publications, Belmont, California, 1984.

[45] R. Rao and H. Toutenburg, *Linear models. Least squares and alternatives,* 2nd ed., Springer, New York, 1999.

[46] W. F. Massy, *Principal components regression in exploratory statistical research*, *Journal of the American statistical Association* 60 (1965) 234-246.

[47] T. Jolliffe, *Principal component analysis*, 2nd ed., Springer, New York, 2002.

[48] http://reference.wolfram.com/mathematica/ref/NonlinearModelFit.html

[49] D. W. Marquardt, Generalized inverses, ridge regression, biased linear estimation, and nonlinear estimation, *Technometrics*, 12(3) (1970) 591-612.

[50] http://www.salfordsystems.com/mars.php.

[51] P. Craven and G. Wahba , *Smoothing noisy data with spline functions: estimating the correct degree of smoothing by the method of generalized cross-validation*, *Numer. Math.* 31 (1979) 377-403.

Chapter 3

PARAMETRIC MODELS OF EFFICIENCY AND OUTPUT POWER OF COOPER BROMIDE LASER

ABSTRACT

In this chapter, we have used classical multivariate parametric models such as: cluster analysis, factor analysis and principal component regression.

These methods are applied in order to find and study the general relationships between the basic laser characteristics (further on referred to as variables) of the investigated copper bromide vapor lasers based on experimental data. The ultimate goal is to construct mathematical models, which describe and predict the results of the experiment.

Such studies in the area of metal vapor lasers have not been conducted up to this point and the developed procedures and obtained results are new.

To be more precise, the goals of the construction of these empirical models are:

- Classifying independent input variables, according to their proximity to output laser characteristics – laser efficiency and output power, and defining the main input variables, reliable for modeling
- Classifying independent variables by grouping those which correlate strongly with each other and with output variables but do not correlate with the rest
- Determining the relative importance of grouped variables (factors) on the output variables

- Setting up multiple linear parametric models of the types of relationships between basis and output quantities
- Proposing methods for predicting the experiment using the constructed models
- Interpreting the results and highlighting the most important physics processes, which influence the efficiency and output power of the laser system.

One of the main objectives in this chapter is to construct empirical regression models, capable of finding linear relationships of type (2.29)-(2.30).

In view of the fact that our input variables are mutually correlated, factor analysis is carried out in advance and the grouping variables (factors) are defined, describing the predominant dimensions of the data cloud. The method applied here is that of principle component regression, described in 2.3.5. More specifically, it is shown how these models can be used to predict the experiment.

The statistical analysis is performed using a representative sample from the available experimental data for copper bromide vapor lasers, published in literature, given in Table 2.1.

All calculations in this Chapter have been carried out using SPSS [1]. The SPSS output plots and tables have been preserved.

3.1. Preliminary Data Analysis

As it was described in 2.1.2, the statistical analyses in this study are conducted on the basis of historical data. A list of the investigated CuBr lasers and basic characteristics are given in Tables 2.1 and 2.2.

From the references, cited in Table 2.1, the data about roughly 300 experiments can be extracted, some of which incomplete and repeated. Also, the number of experiments is very irregularly distributed among the 8 types of lasers. In [2-5] random samples from all available data were used and satisfactory results were obtained.

Another peculiarity of the data is that some variables, such as *PRF, PNE, TR*, and *C*, are not statistically significant and were excluded from some of the analyses. This can be considered as a restrictive condition at which the statistical investigation has been conducted and is only valid when some limitations for these laser characteristics are kept. Therefore, it is impossible to accurately determine the influence of these variables in some of the models.

In this Chapter, we employ a representative sample of size n=121. Where possible, we have avoided the cases with a large number of repeated values, especially of the first 6 variables, which demonstrate the highest reliability. After an analysis, the outliers have also been removed.

Further on, we present the basic results of the initial processing of the input data on the copper bromide vapor laser: descriptive statistics of the variables, plots, histograms.

Figure 3.1 shows the matrix with the scatter plots of all variables. The last two rows are the plots, expressing the relationship between all independent variables and the dependent variables *Pout* – laser generation (output power) and *Eff* – laser efficiency, respectively. The upper rows represent the individual plots for each of the 10 independent input variables. Although more arbitrary – some of them are characterized by a raw general linear trend and others by a more concentrated peaks (nonlinearitites).

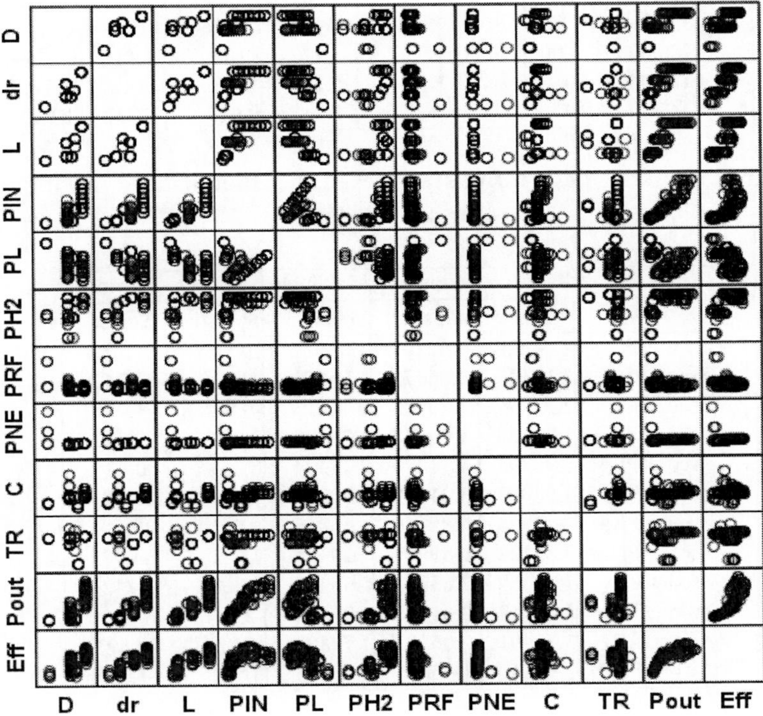

Figure 3.1. Scatterplot matrix of all 12 variables for a representative sample of 121 experiments on the CuBr lasers.

Table 3.1 and 3.2 present a summary of the basic descriptive statistics of unstandardized and standardized variables in the sample, respectively.

Table 3.1. Basic descriptive statistics of the dataset for a CuBr laser

	Range	Minimum	Maximum	Mean	Std. Deviation	Variance	Skewness	Kurtosis
	Statistic	Statistic	Statistic	Statistic	Statistic	Statistic	Statistic	Statistic
D	43.00	15.00	58.00	47.01	10.622	112.823	-0.860	1.134
dr	53.50	4.50	58.00	39.89	17.140	293.765	-0.326	-1.242
L	170.00	30.00	200.00	124.63	67.602	4570.069	-0.020	-1.725
PIN	4.00	1.00	5.00	2.282	1.203	1.447	0.873	-0.382
PL	11.67	5.00	16.67	10.14	2.989	8.936	0.059	-0.830
PH2	0.60	0.00	0.60	0.49	0.152	0.023	-1.393	1.452
PRF	38.00	14.00	52.00	18.50	5.108	26.088	4.861	29.699
PNE	85.00	15.00	100.00	19.83	8.6284	74.449	7.481	64.872
C	3.30	.00	3.30	1.09	0.4132	0.171	1.068	7.742
TR	180.00	350.00	530.00	471.240	33.177	1100.684	-2.417	6.576
Pout	119.00	1.00	120.00	45.19	35.119	1233.340	0.512	-1.105
Eff	2.97	0.10	3.07	1.72	0.783	0.613	-0.232	-0.893

Table 3.2. Basic descriptive statistics of the standardized input data for a CuBr laser

	Range	Minimum	Maximum	Skewness	Kurtosis
	Statistic	Statistic	Statistic	Statistic	Statistic
Zscore(D)	4.04826	-3.02200	1.02627	-0.860	1.134
Zscore(dr)	3.12143	-2.06472	1.05671	-0.326	-1.242
Zscore(L)	2.51471	-1.39978	1.11493	-0.020	-1.725
Zscore(PIN)	3.32488	-1.06547	2.25941	0.873	-0.382
Zscore(PL)	3.90394	-1.71870	2.18524	0.059	-0.830
Zscore(PH2)	3.94035	-3.21090	0.72945	-1.393	1.452
Zscore(PRF)	7.43983	-0.88119	6.55863	4.861	29.699
Zscore(PNE)	9.85122	-0.55937	9.29185	7.481	64.872
Zscore(C)	7.98669	-2.63203	5.35466	1.068	7.742
Zscore(TR)	5.42552	-3.65438	1.77114	-2.417	6.576
Zscore(Eff)	3.79401	-2.06503	1.72899	-0.232	-0.893
Zscore(Pout)	3.38849	-1.25823	2.13026	0.512	-1.105

Figures 3.2 and 3.3 show histograms of the types of the distributions of the dependent variables *Eff*, *Pout* and the logarithmic transformation *LPout* of *Pout*. The results from the formal one-sample Kolmogorov-Smirnov test are

given in Table 3.3. For *Eff* and *LPout* the indices of Asympotic Sig. (2-tailed) and Monte Carlo Sig. (2-tailed) at the 99% level are over 0.05. It can be concluded that the distributions of these variables based on the sample are closed to normal. However, an assumption of slight nonnormality of the general population seems to be more realistic.

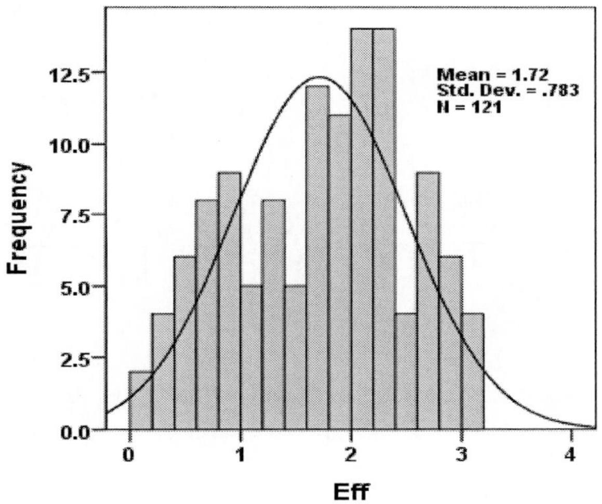

Figure 3.2. Histogram of the dependent variable *Eff*.

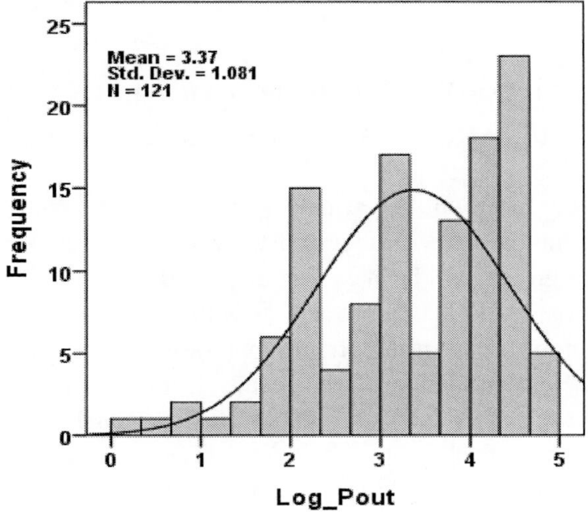

Figure 3.3. Histogram of the dependent variable *Log_Pout*.

Table 3.3. Results form Kolmogorov-Smirnov tests of normality of dependent variables

One-Sample Kolmogorov-Smirnov Test					
			Eff	Pout	Log_Pout
N			121	121	121
Normal Parameters[a,b]	Mean		1.7165	45.1876	3.3723
Kolmogorov-Smirnov Z			1.039	1.922	1.307
Asymp. Sig. (2-tailed)			0.230	0.001	0.066
Monte Carlo Sig. (2-tailed)	Sig.		0.215[c]	0.001[c]	0.058[c]
	99% Confidence Interval	Lower Bound	0.204	0.000	0.052
		Upper Bound	0.226	0.002	0.064

a. Test distribution is Normal.
b. Calculated from data.
c. Based on 10000 sampled tables with starting seed 2000000.

Since the extracted dataset is relatively small, the used sample of 121 experiments gives relatively good representation of the population. Actually, for relatively small samples, at a standard 95% significance level (Sig. = α =0.05), it is recommended that the sample size is above $n > 60$, provided the model accounts for 95% of the total variance (see for instance [6, pp. 221]). We will try to observe this "95%-95%" rule.

3.2. Classification of Variables Using Hierarchical Cluster Analysis

We are going to group the variables for a copper bromide vapor laser in order to determine the inner structure of the investigated dataset. The mutual proximity of the groups will be classified as well as their proximity to the laser output characteristics. This will provide us with a reference cluster model for the classification of the most important input variables and groups, as well as for comparison and validation of the results from the previous and further multivariate models. Obtaining a reliable cluster model will provide good directions for the selection of the basis laser characteristics and their relationships when conducting physical and computer experiments.

The goal is to determine which of the variables are similar (close) in their behavior and what their similarity or dissimilarity is with the investigated dependent variables *Eff* – laser efficiency and *Pout* – output laser power.

We are going to perform an agglomerative hierarchical cluster analysis (CA) of variables which is the most recommendable for samples of size less than 500. The clustering procedures were demonstrated by using the most widely used methods - Average linkage (Between groups) and the Further neighbor (Complete linkage) with the most frequently used metric – squared Euclidean distance. All calculations are based on the standardized variables in order to avoid the influence of the units of measurement (see Chapter 2).

3.2.1. Classification of All Variables

The first step of hierarchical CA is the calculation of the proximity matrix. The obtained results are given in Table 3.4.

We have to note that these are the distances at the first step, when each variable forms one cluster.

In Table 3.4 the smallest distance 16.37 is between variables L and dr. During the first clustering these variables are combined into a first new cluster.

After that, the CA algorithm is employed to calculate the distances from this cluster to all others using the formula for the chosen method (in this case (2.15)); the smallest distance between the clusters is determined and the next grouping is performed, and so on.

All clusters, obtained using the Average linkage method with clustership of the variables are shown in Table 3.5.

The second step is to determine the number of clusters in the solution. There are several different mathematical procedures for calculating the number of clusters but they are complex (and controversial) and are not applicable (see [7], page 494 and the references there). In practice, the so-called dendrogram is used to solve the problem.

First, we will attempt to find the number of clusters for our model by applying the Average linkage (Between groups) method. The obtained dendrogram is shown in Figure 3.4. The names of the variables in clustering procedure are given in the first column. The second column contains their corresponding numbers in the dataset. The scale of measurement always uses ratio scales with a standard of 25 units, where the maximum length of 25 corresponds to the largest distance in the proximity matrix (in our case 391.74).

Table 3.4. Proximity matrix for the 10 standardized variables from the data set for a copper bromide vapor laser with a squared Euclidean distance

Proximity Matrix										
Variable	Matrix File Input									
	D	dr	L	PIN	PL	PH2	PRF	PNE	C	TR
D	0.00	39.12	37.80	61.53	363.78	145.64	356.19	320.54	165.55	192.70
dr	39.12	0.00	16.37	50.60	379.43	68.61	308.80	264.50	173.51	186.81
L	37.80	16.37	0.00	39.26	391.74	91.17	309.32	246.46	193.24	192.91
PIN	61.53	50.60	39.26	0.00	279.52	116.96	297.80	243.02	170.52	195.07
PL	363.78	379.43	391.74	279.52	0.00	344.54	159.06	191.65	243.87	237.40
PH2	145.64	68.61	91.17	116.96	344.54	0.00	258.77	222.92	249.24	284.79
PRF	356.19	308.80	309.32	297.80	159.06	258.77	0.00	178.71	293.48	242.78
PNE	320.54	264.50	246.46	243.02	191.65	222.92	178.71	0.00	294.326	235.21
C	165.55	173.51	193.24	170.52	243.87	249.24	293.48	294.32	0.00	107.18
TR	192.7	186.81	192.91	195.07	237.40	284.79	242.78	235.21	107.18	0.00

Table 3.5. Cluster membership of the 10 independent variables using the method of Average linkage (Between groups) and Squared Euclidean measure

Cluster Membership								
Case	9 Clusters	8 Clusters	7 Clusters	6 Clusters	5 Clusters	4 Clusters	3 Clusters	2 Clusters
D	1	1	1	1	1	1	1	1
dr	2	1	1	1	1	1	1	1
L	2	1	1	1	1	1	1	1
PIN	3	2	1	1	1	1	1	1
PL	4	3	2	2	2	2	2	2
PH2	5	4	3	1	1	1	1	1
PRF	6	5	4	3	3	2	2	2
PNE	7	6	5	4	4	3	2	2
C	8	7	6	5	5	4	3	1
TR	9	8	7	6	5	4	3	1

Figure 3.4 shows that variables *DR*, *L*, *D* and *PIN* are very closed and were grouped first. Then *PH2* was added to this group at a much greater distance. Other similar group is *C* and *TR*, then *PL*, *PRF* and *PNE*. Usually, the number of clusters for the best possible solution is chosen when there is a gap of at least 5 relative units.

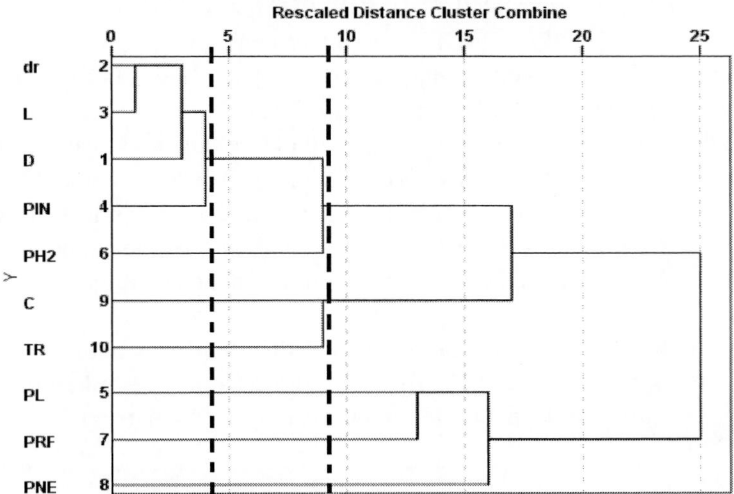

Figure 3.4. Dendrogram of all ten input variables, obtained by the Average linkage (Between groups) clustering method and squared Euclidean distance.

Normally, the formation of one- and two-cluster solutions are not advisible due to the excessively long distances. It is seen that the only one gap with at least 5 units is in the interval [0,10]. In order to facilitate the analysis we have plotted 2 additional vertical lines in the dendrogram (namely the bold dotted lines), clearly identifying this largest gap into clusters. The right line crosses 5 horizontal lines and corresponds to the solution with 5 clusters (the same with the 6), exactly identified in the 5-cluster membership in Tabl. 3.5. These are: {*DR*, *L*, *D*, *PIN*, *PH*2}, {*PL*}, {*PRF*}, {*PNE*} and {*C*, *TR*}. The left bold dotted line croses 7 horizontal lines and so corresponds to seven clusters. This way, with the help of the dendrogram in Figure 3.4, we can assume that formally the best solution at this stage is that with 7 clusters (compare with Table 3.5):

Cluster 1: {*DR*, *L*, *D*, *PIN*}
Cluster 2: {*PH*2}
Cluster 3: {*PL*} (3.1)
Cluster 4: {*PRF*}
Cluster 5: {*PNE*}
Cluster 6: {*C*}
Cluster 7: {*TR*}

However, following the formal clustering procedure in Figure 3.4 we can distinguish a cluster group {*DR*, *L*, *D*, *PIN*, *PH2*}, a second group {*C*, *TR*} and a third group {*PL*, *PRF*, *PNE*}. We can also add here that Figure 3.4 is qualitatively identical with Figure 2 from [2] for the same method and distance, obtained for a sample containing 74 cases.

In addition to the Average linkage method, calculations were performed using six other methods with various distances. The predominant resulting solution is that with 3 clusters, obtained using squared Euclidean, Euclidean and Minkowski measures (see Chapter 2). Of these experiments, here we have presented the solution, reached using the Further linkage method.

Table 3.6. Cluster membership of the 10 input variables using the method of Further neighbor (Complete linkage) and Squared Euclidean measure (2 to 7 clusters)

Case	7 Clusters	6 Clusters	5 Clusters	4 Clusters	3 Clusters	2 Clusters
D	1	1	1	1	1	1
dr	1	1	1	1	1	1
L	1	1	1	1	1	1
PIN	1	1	1	1	1	1
PL	2	2	2	2	2	2
PH2	3	3	1	1	1	1
PRF	4	4	3	2	2	2
PNE	5	5	4	3	2	2
C	6	6	5	4	3	1
TR	7	6	5	4	3	1

Now, the only big gap between the clusters (excluding the first two) is in the interval [10, 20]. The left bold dotted line intersects 3 horizontal lines. The resulting 3 clusters have structures similar to those shown in Figure 3.4, without any unnecessary fragmentation of the groups.

It can be concluded that the grouping and the structure of the distribution of the variables within the clusters are best described by the following 3-cluster solution:

Cluster 1: {*DR*, *L*, *D*, *PIN*, *PH2*}
Cluster 2: {*PL*, *PRF*, *PNE*} (3.2)
Cluster 3: {*C*, *TR*}

It has to be noted that in Figure 3.4 and 3.5 within the first cluster the variable *PH2* is at a distance from the other four variables of more than 5 relative units. The type of dissimilarity between *PL* and {*PRF*, *PNE*}, is similar to that between *C* and *TR*.

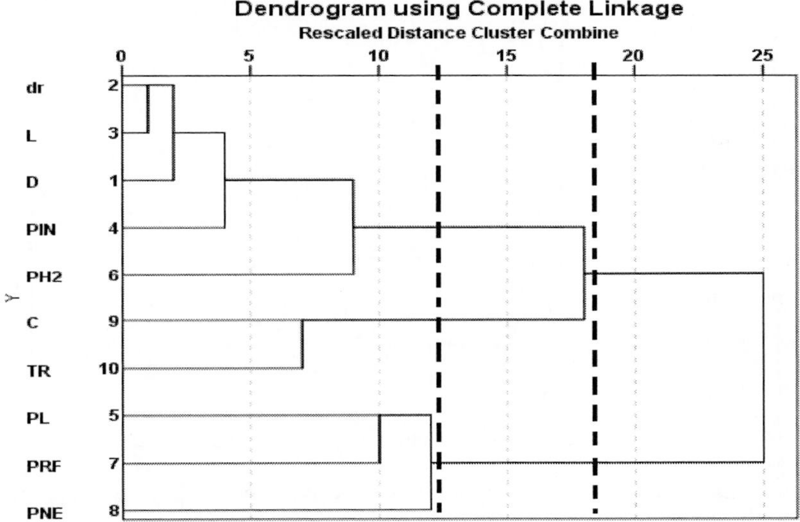

Figure 3.5. Dendrogram of all ten input variables, obtained by the Further neighbor (Complete linkage) clustering method and squared Euclidean distance.

The next step is to classify the 10 input variables together with the dependent ones by consecutively carrying out CA of the mentioned variables with *Eff* and then with *LPout*. The dendrogram with *Eff* is given in Figure 3.6, and the dendrogram with *LPout* - in Figure 3.7. For the sake of simplicity, we have presented only the results, obtained using the Further neighbor (Complete) method and squared Euclidean distance.

The dendrograms clearly show that *Eff* is grouped with the variables of the first cluster in (3.2) - *dr*, *L*, *Pin*, *D*, *PH2*. The same is true for *LPout*. Therefore, these variables show the greatest degree of similarity with the dependent variables, while the remaining variables *C*, *PRF*, *PNE*, *TR* and *PL* are significantly different. The specific strength, type and direction of this will be futher examining using other statistical techniques.

The results, obtained using other distances and methods have not been described here, in order not to overburden the presentation. But as a whole they include the same groups as (3.2), although not always at a great distance. At this stage, it can be concluded that the general structure within the classifications of the three basis groups is always maintained.

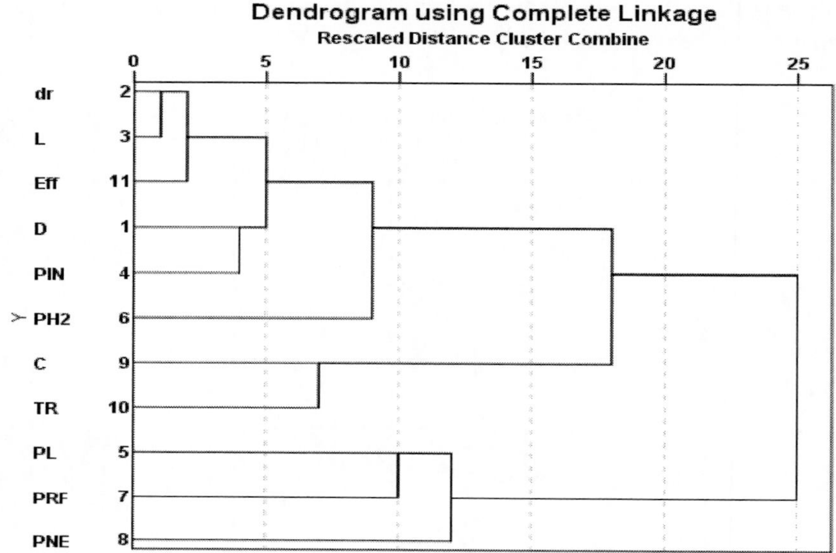

Figure 3.6. Dendrogram of 10 input variables and dependent variable *Eff* using Further neighbor (Complete linkage) and squared Euclidean distance.

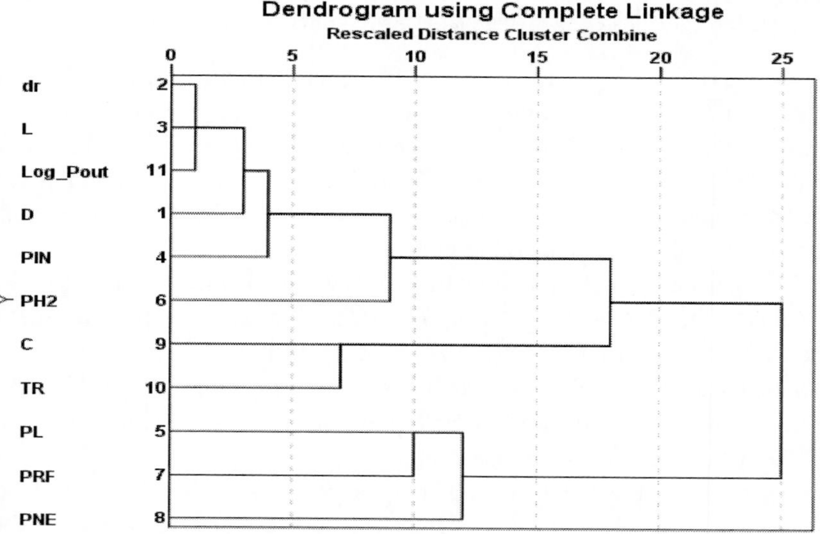

Figure 3.7. Dendrogram of 10 input variables and dependent variable *Log_Pout* (*LPout*) using Further neighbor (Complete linkage) and squared Euclidean distance.

3.2.2. Classification of Six Basic Variables

For the purpose of further analyses, we are going to present the results of the classification only of six basic variables – *D*, *dr*, *L*, *PIN*, *PL* and *PH2*, obtained using the Complete linkage method with the squared Euclidean distance.

The proximity matrix is given in Table 3.7. In this case, it is a sub-matrix of the 10-variable matrix from Table 3.4.

Table 3.7. Proximity matrix for the 6 standardized variables from the data set for a copper bromide vapor laser with a squared Euclidean distance

Proximity Matrix						
	Matrix File Input					
Case	D	dr	L	PIN	PL	PH2
D	0.000	39.121	37.804	61.527	363.780	145.635
dr	39.121	0.000	16.369	50.599	379.434	68.611
L	37.804	16.369	0.000	39.255	391.738	91.173
PIN	61.527	50.599	39.255	0.000	279.520	116.962
PL	363.780	379.434	391.738	279.520	0.000	344.541
PH2	145.635	68.611	91.173	116.962	344.541	0.000

The dendrogram is given in Figure 3.8. It is obvious to conclude that the most suitable grouping is in three clusters. The corresponding cluster models are equivalent and the solution is given in Table 3.8.

Table 3.8. Cluster membership of the 6 basic input variables using the method of Further neighbor (Complete linkage) and Squared Euclidean measure

Case	5 Clusters	4 Clusters	3 Clusters	2 Clusters
D	1	1	1	1
dr	2	1	1	1
L	2	1	1	1
PIN	3	2	1	1
PL	4	3	2	2
PH2	5	4	3	1

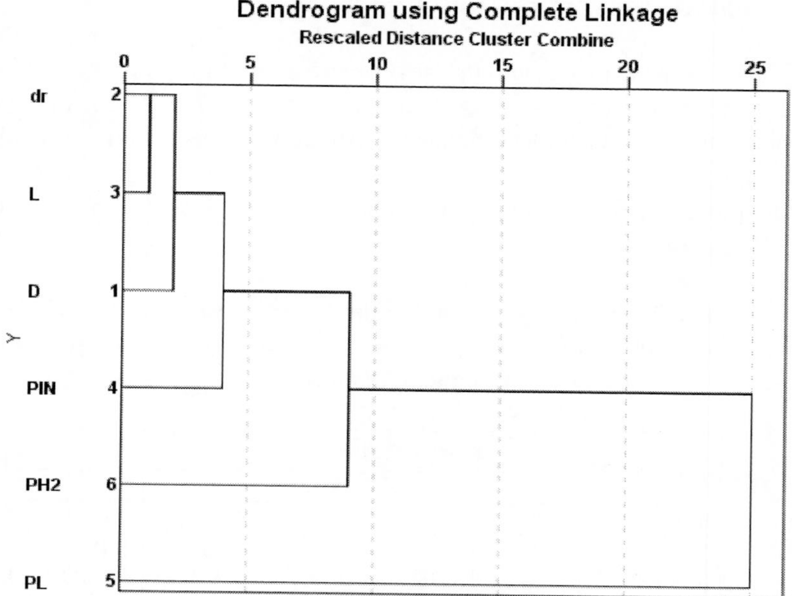

Figure 3.8. Dendrogram of the clusters for 6 basic input variables obtained using the method for Complete linkage with Squared Euclidean distance.

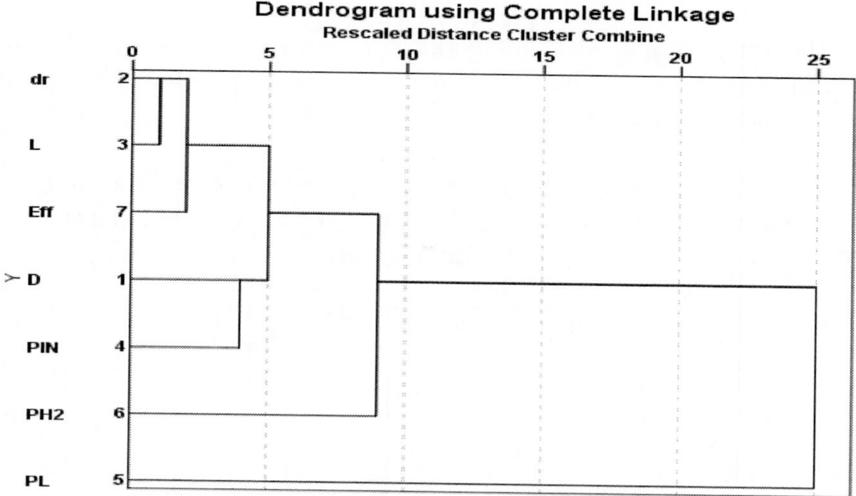

Figure 3.9. Dendrogram of the clusters for 6 basic input variables and *Eff* obtained using the method for Complete linkage with Squared Euclidean distance.

The final clustering procedure is to classify the 6 basic input variables together with the dependent variables. The dendrogram with *Eff* is given in Figure 3.9, and the dendrogram with *LPout* - in Figure 3.10. For the sake of simplicity, we have presented only the results, obtained using the Further neighbor (Complete) method and squared Euclidean distance. We observe the same type of structures as in the case of ten input variables with the dependent variables.

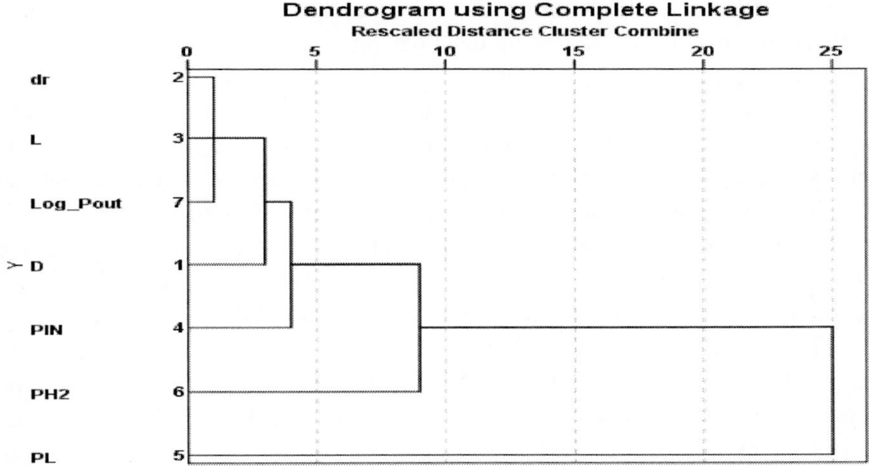

Figure 3.10. Dendrogram of the clusters for 6 basic input variables and *Log_Pout* obtained using the method for Complete linkage with Squared Euclidean distance.

3.3. Application of Multiple Factor Analysis for a CuBr Laser

This section illustrates the results of the FA performed with the goal of constructing further parametric and nonparametric regression models of output laser characteristics. This technique is applied to reveal the relationships between independent laser variables and to take into account of their mutual collinearity. As a result, the dimensions of the initial dataset were reduced.

FA was performed in accordance with the procedures described in 2.3.3. The necessary basic statistical tests are performed at each step.

In the beginning, the very simplified results based on a small matched sample of only seven experiments for six laser variables were obtained. A factor model and a regression model with two factors [8] were constructed. Another factor models for a copper bromide vapor laser have been published in [3, 4, 9, 10].

3.3.1. Testing Basic Assumptions for Factor Analysis

We continue to use the sample of size $n = 121$, investigated in 3.1.

In accordance with section 2.3.3, we will check if the necessary theoretical assumptions are fulfilled, allowing us to perform FA.

The used experimental data are of interval type.

Another requirement is that the number of observations is substantially higher than the number of input independent variables. In this case the variables are $p = 10$. The ratio $n / p = 12$ is satisfactory. Thus, the dataset contains enough observations for the analysis to be performed.

Our next task is to determine which variables correlate and demonstrate reliability so as to include these in the FA procedures. To this end, we calculate the correlation matrix, also including the dependent variables.

The general correlation matrix of the standardized variables is given in Table 3.9. For comparison, the variable *Pout* is also included, but will not be used in analysis. The bivariate correlation coefficients are given in the upper half of the table, and their respective one-tailed statistical significance (Sig.) – in the lower half. As noted previously, all investigations have been conducted at the $\alpha = 0.05$ significance level.

The first six variables *D*, *DR*, *L*, *PIN*, *PL* and *PH*2 demonstrate high correlation coefficients with *Eff* and *LPout.* Their respective values from the last rows of the table are Sig. < 0.05, so they are statistically significant. In addition, these six variables correlate relatively high with each other.

The remaining four variables *PRF*, *PNE*, *C* and *TR* do not correlate with *Eff* and *LPout.* This lack of correlation is significant in a half of cases. Moreover, these four variables do not correlate with the remaining ones, although a part of the corresponding Sig. coefficients show their statistical insignificance. In accordance with the rules for FA, these variables are inadequate at this stage and should be excluded from the analysis as unrealiable.

The analysis of the correlation matrix and the almost zero value of the determinant lead to the conclusion that there is a case of multicolinearity between the variables involved. What is more, in subsequent investigations, only the first six input variables *D*, *DR*, *L*, *PIN*, *PL* and *PH*2 may be used.

Table 3.9. Correlation matrix of all 13 variables

		Correlation Matrix[a]												
		D	DR	L	PIN	PL	PH2	PRF	PNE	C	TR	Eff	Pout	LPout
Correlation	D	1.000	0.837	0.842	0.744	-0.516	0.393	-0.484	-0.336	0.310	0.197	0.726	0.791	0.799
	DR	0.837	1.000	0.932	0.789	-0.581	0.714	-0.287	-0.102	0.277	0.222	0.875	0.864	0.907
	L	0.842	0.932	1.000	0.836	-0.632	0.620	-0.289	-0.027	0.195	0.196	0.846	0.897	0.884
	PIN	0.744	0.789	0.836	1.000	-0.165	0.513	-0.241	-0.013	0.289	0.187	0.663	0.939	0.823
	PL	-0.516	-0.581	-0.632	-0.165	1.000	-0.436	0.337	0.201	-0.016	0.011	-0.590	-0.311	-0.496
	PH2	0.393	0.714	0.620	0.513	-0.436	1.000	-0.078	0.071	-0.039	-0.187	0.707	0.570	0.734
	PRF	-0.484	-0.287	-0.289	-0.241	0.337	-0.078	1.000	0.255	-0.223	-0.012	-0.232	-0.245	-0.268
	PNE	-0.336	-0.102	-0.027	-0.013	0.201	0.071	0.255	1.000	-0.226	0.020	-0.100	-0.020	-0.207
	C	0.310	0.277	0.195	0.289	-0.016	-0.039	-0.223	-0.226	1.000	0.553	0.095	0.232	0.204
	TR	0.197	0.222	0.196	0.187	0.011	-0.187	-0.012	0.020	0.553	1.000	0.044	0.202	0.047
	Eff	0.726	0.875	0.846	0.663	-0.590	0.707	-0.232	-0.100	0.095	0.044	1.000	0.848	0.931
	Pout	0.791	0.864	0.897	0.939	-0.311	0.570	-0.245	-0.020	0.232	0.202	0.848	1.000	0.904
	LPout	0.799	0.907	0.884	0.823	-0.496	0.734	-0.268	-0.207	0.204	0.047	0.931	0.904	1.000
Sig. (1-tailed)	D		0.000	0.000	0.000	0.000	0.000	0.000	0.000	0.000	0.015	0.000	0.000	0.000
	DR	0.000		0.000	0.000	0.000	0.000	0.001	0.133	0.001	0.007	0.000	0.000	0.000
	L	0.000	0.000		0.000	0.000	0.000	0.001	0.385	0.016	0.016	0.000	0.000	0.000
	PIN	0.000	0.000	0.000		0.036	0.000	0.004	0.445	0.001	0.020	0.000	0.000	0.000
	PL	0.000	0.000	0.000	0.036		0.000	0.000	0.013	0.430	0.453	0.000	0.000	0.000
	PH2	0.000	0.000	0.000	0.000	0.000		0.197	0.219	0.337	0.020	0.000	0.000	0.000
	PRF	0.000	0.001	0.001	0.004	0.000	0.197		0.002	0.007	0.450	0.005	0.003	0.001
	PNE	0.000	0.133	0.385	0.445	0.013	0.219	0.002		0.006	0.414	0.137	0.415	0.011
	C	0.000	0.001	0.016	0.001	0.430	0.337	0.007	0.006		0.000	0.151	0.005	0.012
	TR	0.015	0.007	0.016	0.020	0.453	0.020	0.450	0.414	0.000		0.316	0.013	0.304
	Eff	0.000	0.000	0.000	0.000	0.000	0.000	0.005	0.137	0.151	0.316		0.000	0.000
	Pout	0.000	0.000	0.000	0.000	0.000	0.000	0.003	0.415	0.005	0.013	0.000		0.000
	LPout	0.000	0.000	0.000	0.000	0.000	0.000	0.001	0.011	0.012	0.304	0.000	0.000	

[a]. Determinant = 2.64E-009.

The next important stage is to determine the meaninfullness of performing a factor analysys. The capabilities of the SPSS software package are used to calculate the Kaiser-Meyer-Olkin (KMO) measure of sampling adequacy and Bartlett's test of sphericity. These statistics provide more exact information whether a factor analysis is significant or not. In Table 3.10 for the ten independent variables KMO =0.0.575>0.5 and Bartlett's test is statistically significant with Sig. =0.000 < 0.05. The conclusion is that FA is formally adequate and can be performed using this sample with all 10 variables. However, the examination of the correlation matrix shows that this is not the case.

Table 3.10. Adequacy test for FA with 10 variables for a CuBr laser

KMO and Bartlett's Test		
Kaiser-Meyer-Olkin Measure of Sampling Adequacy.		0.575
Bartlett's Test of Sphericity	Approx. Chi-Square	1147.744
	df	45
	Sig.	0.000

Table 3.11. Correlation matrix of the six variables selected

Correlation Matrix[a]							
		D	DR	L	PIN	PL	PH2
Correlation	D	1.000	0.837	0.842	0.744	-0.516	0.393
	DR	0.837	1.000	0.932	0.789	-0.581	0.714
	L	0.842	0.932	1.000	0.836	-0.632	0.620
	PIN	0.744	0.789	0.836	1.000	-0.165	0.513
	PL	-0.516	-0.581	-0.632	-0.165	1.000	-0.436
	PH2	0.393	0.714	0.620	0.513	-0.436	1.000
Sig. (1-tailed)	D		0.000	0.000	0.000	0.000	0.000
	DR	0.000		0.000	0.000	0.000	0.000
	L	0.000	0.000		0.000	0.000	0.000
	PIN	0.000	0.000	0.000		0.036	0.000
	PL	0.000	0.000	0.000	0.036		0.000
	PH2	0.000	0.000	0.000	0.000	0.000	

[a]. Determinant =0 .001.

In addition, a separate correlation matrix of the six independent variables is also presented in Table 3.11. This matrix is a submatrix of the common correlation matrix in Table 3.9 with a determinant of 0.001. The adequacy test indicates that FA can be performed using only these variables (see also Table 2.12). The actual obtained values are: KMO=0.658 >0.5 and Sig. =0.000, which leads to the conclusion that FA is a good idea.

Table 3.12. Adequacy test of the selected six basic variables

KMO and Bartlett's Test		
Kaiser-Meyer-Olkin Measure of Sampling Adequacy.		0.658
Bartlett's Test of Sphericity	Approx. Chi-Square	880.999
	df	15
	Sig.	0.000

3.3.2. Performing Factor Analysis for a CuBr Laser Dataset Based on Six Basic Variables

During the next step of FA, the artificial variables (factors) are extracted and the distribution of variance is calculated. The techniques and methods are briefly described in 2.3.3. We will use the method of Principal component analysis (PCA) to extract factors. In this method the obtained factors explain a maximum possible share of total variance in data. The first factor explains the largest possible part of the total variance. The second factor explains the largest possible part of the remaining total variance, and so on. The obtained variables are orthogonal and so uncorrelated to each other.

In our case, PCA was applied to extract the six possible factors (components). The part of the variance associated with any component is expressed by the eigenvalues of the correlation matrix. These eigenvalues for our data sample are shown in the column 'Initial Eigenvalues', sub-column 'Total' in Table 3.13. They are presented in descending order, according to the corresponding variance in the total explained variance. In this case, the sum of the eigenvalues is 6, which is equal to the number of the variables involved in the analysis. The percent of variance, explained from the factors is given in the next column '% of Variance'. The next column to the right 'Cumulative %" contains the accumulated percent.

Table 3.13. Variance distribution for FA with six variables

Total Variance Explained									
	Initial Eigenvalues			Extraction Sums of Squared Loadings			Rotation Sums of Squared Loadings		
Component	Total	% of Variance	Cumulative %	Total	% of Variance	Cumulative %	Total	% of Variance	Cumulative %
1	4.268	71.128	71.128	4.268	71.128	71.128	2.968	49.472	49.472
2	0.855	14.247	85.374	0.855	14.247	85.374	1.419	23.651	73.123
3	0.632	10.531	95.905	0.632	10.531	95.905	1.367	22.782	95.905
4	0.155	2.581	98.486						
5	0.069	1.145	99.631						
6	0.022	0.369	100.000						

Extraction Method: Principal Component Analysis.

Table 3.13 is used to determine the number of factors for our model. In theory, when working with a large number of variables (20-50 or more), it is recommended that the number of factors is equal to the number of the eigenvalues of the correlation matrix, which have values over or equal to 1. The table shows that one factor accounts for only 71.128% of the total variance. For two factors with accumulation, the factor model accounts for 85.374% of the variance. If three factors are taken, the accumulated percentage is significantly higher: 95.905%.

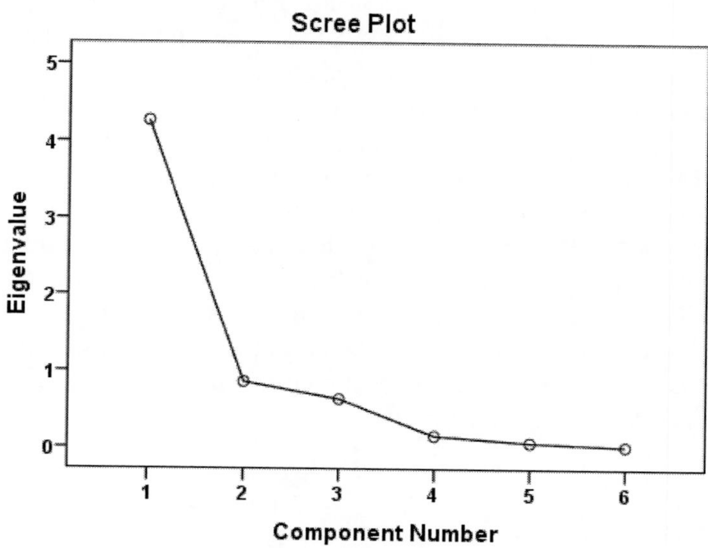

Figure 3.11. Scree plot for choosing the number of factors in the model.

The graphic representation of the eigenvalues using the so-called scree plot is often useful when choosing the number of factors. The scree plot for the problem at hand is given in Figure 3.11

According to the scree plot, it is recommendable that the chosen number of factors is equal to the last point at which there is a steep drop in the values. In our case, 2 or 3 factors may be chosen. We have chosen 3 factors, due to the relatively small sample size (n=121) and since we are working at the 0.05 significance level, the samples should provide a high percentage share when accounting for analyses. Another motive is that given the strong relationships between the variables of physics processes, it is better to find a clearer differentiation within the structure of these relationships. The issue of choosing the best number of factors for MVLs and the point of view presented in statistics literature in this direction is discussed in detail in 2.3.3, FA using PCA.

Another confirmation that the chosen number of three factors is correct is the result from the CA, performed in section 3.2, Table 3.6, where the same solution was obtained.

Based on these arguments, we have chosen to construct a factor model with three factors.

Table 3.14. Matrix of the reproduced correlations and their residuals

Reproduced Correlations							
		D	DR	L	PIN	PL	PH2
Reproduced Correlation	D	0.920[a]	0.838	0.889	0.783	-0.539	0.361
	DR	0.838	0.947[a]	0.944	0.816	-0.597	0.723
	L	0.889	0.944	0.958[a]	0.817	-0.622	0.637
	PIN	0.783	0.816	0.817	0.959[a]	0-.140	0.524
	PL	-0.539	-0.597	-0.622	-0.140	0.985[a]	-0.442
	PH2	0.361	0.723	0.637	0.524	-0.442	0.985[a]
Residual[b]	D		-0.001	-0.046	-0.040	0.023	0.033
	DR	-0.001		-0.012	-0.027	0.016	-0.009
	L	-0.046	-0.012		0.020	-0.011	-0.017
	PIN	-0.040	-0.027	0.020		-0.025	-0.012
	PL	0.023	0.016	-0.011	-0.025		0.007
	PH2	0.033	-0.009	-0.017	-0.012	0.007	

Extraction Method: Principal Component Analysis.

[a] Reproduced communalities.

[b] Residuals are computed between observed and reproduced correlations. There are 0 (.0%) nonredundant residuals with absolute values greater than 0.05.

The choice of three factors is also validated by the calculated reproduced correlation matrix and in particular that of the residuals. These are given in Table 3.14. The lower half of the table shows that all residuals have sufficiently small absolute values (under 0.05) which is a very good result. This allows us to conclude that the choice to use three factors for the data is sufficiently appropriate.

The next step of FA is the extraction of the three factors, in accordance with the respective procedures in the Algorithm in FA, described in 2.3.3.5. The initial solution is represented in Table 3.15. The loadings less than 0.5 are omitted in accordance with the restrictions in Table 2.4. It can be noted that the first three variables *L*, *dr* and *D* load highly with the first component (factor) but this is not the case for the last two.

Table 3.15. Initial factor model with six variables, obtained using the Principal Component Analysis

Component Matrix[a]			
	Component		
	1	2	3
L	0.974		
DR	0.972		
D	0.876		
PIN	0.828	0.522	
PH2	0.718		0.668
PL	-0.637	0.736	

Extraction Method: Principal Component Analysis.
[a] 3 components extracted.

In order to obtain a better differentiation of the variables, the factors are rotated during the next step. SPSS offers five factor rotation methods: three methods with orthogonal axes: Varimax, Quartimax, Equamax and two methods with oblique axes: Direct Oblimin, and Promax. Here we present two rotation matrices, obtained using the most common methods: Varimax and Direct Oblimin.

The rotated matrix using Varimax with the loadings higher than 0.5 is given in the left hand side of Table 3.16. Only for comparison, the solution using Direct Oblimin method with loadings higher than 0.4 is shown in right hand side of the same table (see also 2.3.3.4).

Table 3.16. Rotation matrices with factor loadings for the six basic variables, obtained using Varimax and Oblimin rotations

	Component				Component		
	1	2	3		1	2	3
PIN	0.919			PIN	0.964		
D	0.879			D	0.956		
L	0.811			L	0.790		
DR	0.765			DR	0.709		
PL		-0.952		PL		0.927	
PH2			0.932	PH2			0.959
Extraction Method: Principal Component Analysis.				Extraction Method: Principal Component Analysis.			
Rotation Method: Varimax with Kaiser Normalization.				Rotation Method: Oblimin with Kaiser Normalization.			
a. Rotation converged in 5 iterations				a. Rotation converged in 6 iterations			

In this way, a relatively good factor model is constructed. The first four variables *PIN*, *DR*, *L* and *D* are strongly influenced by the first factor and more weakly by the other two. The respective factor loadings are given in columns '1'. The variable *PL* is influenced negatively and most strongly by the second factor. The variable *PH2* is strongly influenced by the third factor. The respective percentages which account for the variance of the sample after the rotation can be seen in the last two columns of Table 3.13, section 'Rotation Sums of Squared Loadings'.

In the end, an acceptable solution with three factors $F1$, $F2$, $F3$ is obtained in the following form:

$$F1 = \{PIN, DR, L, D\}$$
$$F2 = \{PL\} \qquad (3.3)$$
$$F3 = \{PH2\}$$

The three-factor solution from Table (3.16) using Varimax method is represented in Figure 3.12. The rotated solution using Direct Oblimin is shown in Figure 3.13.

The next stage of FA is the calculation of factor scores for each observation and their storage as factor variables for further analyses. An important point in this procedure is that factor variables can also be calculated in the cases where the value of some participating variable is missing. This can be done using different methods, which are not going to be discussed here.

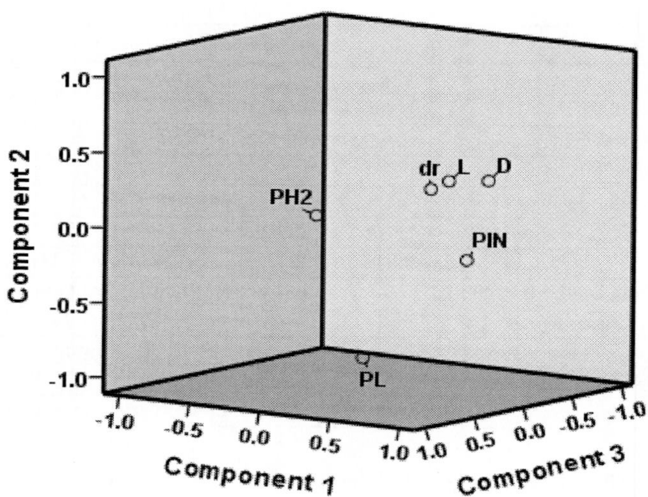

Figure 3.12. Representation of a three-factor model with orthogonal factors, obtained by the Varimax method.

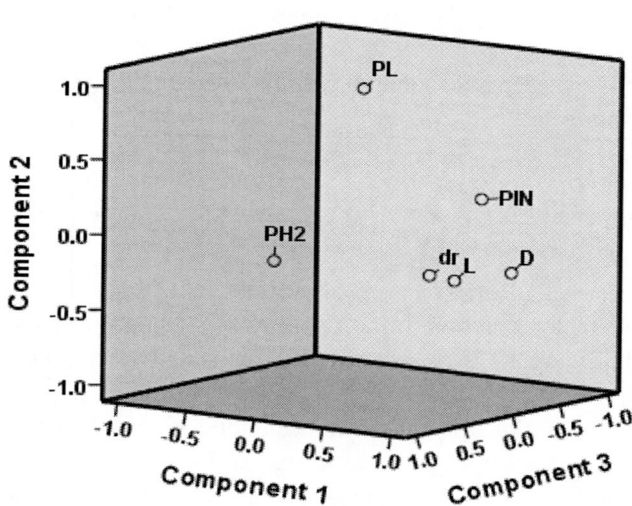

Figure 3.13. Representation of a three-factor model with oblique factors, obtained by the Oblimin method.

The missing values in our data are very few and do not influence the results. The same is the situation with the outliers in the sample.

During the last step of the FA, the factors are named and their most appropriate interpretation is sought. The dominating factor *F*1 can be designated as geometrically-energetic, since it combines the geometric size of the laser tube and input electric power. Factors *F*2 and *F*3 contain one variable each and they can be conveniently called 'factored *PL'* and 'factored *PH*2". In this investigation, we are going to use their mathematical notations *F*1, *F*2, *F*3.

3.4. Application of Principal Component Regression for Modeling Output Characteristics of a CuBr Laser

As noted in Chapter 2, in case of multicollinearity, MLR (Multiple Linear Regression) cannot be applied directly or it provides unsatisfactory results. Alternative methods are employed, such as the one used here, namely the Principal Component Regression (PCR) method. In practice, within this method, the strongly correlated independent variables are transformed into factor variables through FA. The unique variables, which correlate strongly with the dependent variable, are added. These variables and the factor variables are used as predictors when performing multiple linear regression.

We will apply PCR in order to build regression models of the laser output characteristics of a CuBr laser. As shown in 3.1-3.2, there are no unique variables and RA can be conducted with the help of the three factors *F*1, *F*2, *F*3.

3.4.1. Linear Regression Model of Laser Efficiency

The generated factor scores of F1, F2 and F3 are used to perform linear regression analysis with the goal of setting up an explicit model formula for the relationship between the factors and *Eff*. To this end, three different methods for determining the regression coefficients were applied, namely: simple linear method, stepwise regression, and backward linear regression. As it turns out, the models obtained using the three models are the same.

The analysis was carried out in accordance with the procedures, described in 2.3.4

The model summary statistics are given in Table 3.17. The multiple regression coefficient of the model is R = 0.887, and the coefficient of determination R^2=0.786 (see section 2.3.4.3). The latter means the model accounts for 78.6% of the total variance of the sample. This value is relatively high and not very closed to the previously obtained value of 72.2% for another sample in [3]. Another important characteristic is the standard error of the estimate, which is 0.36655, and shows about 12% relative error.

Table 3.17. Summary of the linear regression model for the efficiency of a CuBr laser

Model Summary[b]				
Model	R	R Square	Adjusted R Square	Std. Error of the Estimate
1	0.887[a]	0.786	0.781	0.36655

a. Predictors: (Constant), *F*3, *F*2, *F*1.
b. Dependent Variable: *Eff*.

The ANOVA table of the model is given in Table 3.18. The last column shows the goodness of fit of the model, depending on the value of Sig. If Sig. <0.05, then the model is significant at 95% level, which is satisfied in our case.

Table 3.18. ANOVA table of a linear regression model for the efficiency of a CuBr laser

ANOVA[b]						
Model		Sum of Squares	df	Mean Square	F	Sig.
1	Regression	57.815	3	19.272	143.430	0.000[a]
	Residual	15.720	117	0.134		
	Total	73.535	120			

a. Predictors: (Constant), *F*3, *F*2, *F*1.
b. Dependent Variable: *Eff*.

The obtained coefficients of the model are presented in Table 3.19. The constant and the variables used in the analysis are given in the first column. The calculated non-standardized coefficients $b_0, b_1, \ldots$ (given in the column Beta. The coefficients have been obtained with a satisfactory standard error of 0.033. The t-statistics of the coefficients and their statistical significance are also given, and it can be seen that all coefficients are statistically significant, as their corresponding Sig. = 0.000.

Table 3.19. Coefficients of the linear regression model for the efficiency of a CuBr laser

Coefficients[a]						
		Unstandardized Coefficients		Standardized Coefficients		
Model		B	Std. Error	Beta	t	Sig.
1	(Constant)	1.717	0.033		51.512	0.000
	F1	0.484	0.033	0.618	14.454	0.000
	F2	0.316	0.033	0.404	9.448	0.000
	F3	0.385	0.033	0.491	11.493	0.000

[a] Dependent Variable: *Eff*.

With the help of the calculated coefficients from Table 3.19, the linear regression model of the laser efficiency *Eff* can be represented in the following form:

$$\hat{Eff} = 1.717 + 0.484 F_1 + 0.316 F_2 + 0.385 F_3 \quad (3.4)$$

For the standardized values, the model is:

$$\hat{zEff} = 0.618 F_1 + 0.404 F_2 + 0.491 F_3 \quad (3.5)$$

The obtained regression coefficients in (3.4) and (3.5) show the importance of the three factors (respectively of the grouped variables) on the efficiency. All signs in the models and of the factor levels of the variables for the first and third factor are positive, as per (3.4) and the corresponding values on the left side of Table (3.16). This means that an increase in the value of each variable raises the value of the respective factor and therefore of *Eff*. The variable *PL*, which is the basic one in the second factor has a negative loading (-0.952), indicating that it is inversely proportional to *Eff*.

Equation (3.5) is significant in that it shows the relative strength of influence of the factors on *Eff* and in relation to each other. This shows that the first factor *F*1, which groups together *PIN*, *DR*, *L* and *D*, exerts the strongest influence, and the other two factors *F*2 and *F*3 are almost similar in their influence with a slight prevalence of *F*3.

Figure 3.14 presents the scatter plot of the dependent variable compared to the predicted values of laser efficiency *Eff* with the 95% confidence interval.

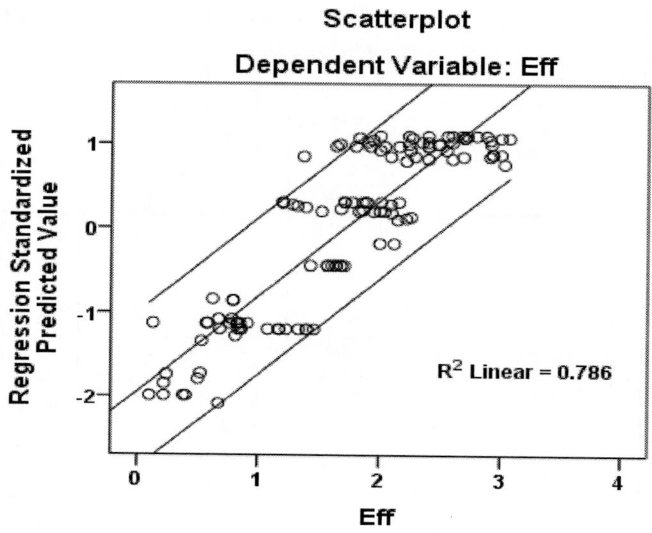

Figure 3.14. Observations versus standardized predicted values of *Eff* in model (3.5) for a copper bromide vapor laser.

The analysis of the residuals is quite essential for each regression model. As mentioned above in 2.3.4, the residuals should be normally distributed. Figure 3.15 displays a histogram of the regressaion standardized residuals. Visually the distribution is closed to normal.

The normal probability plot is also useful. A close fit between the dotted line and the 45 degree-curve is necessary to guarantee normality. Figure 3.16 shows small deviation from the diagonal and also confirms the normality of the residuals.

The standardized residuals and the standardized predicted values are compared in Figure 3.17. We are going to check the assumption of homoscedasticity, which means that the residual must have the same variance for every value of the independent variable. The presence of a strong pattern in the graph can indicate that homoscedasticity does not apply. In our case in Figure 3.17 a weak pattern, composed of lines is observed. To discover which variable adds the elements of heteroscedasticity, we examine the graphs of all independent variables versus residuals.

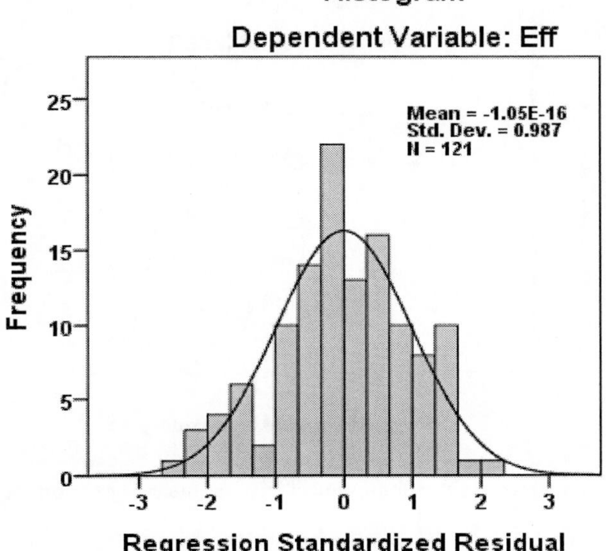

Figure 3.15. Normality test of the residuals from linear regression model (3.4), (3.5).

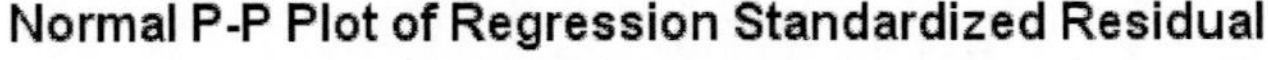

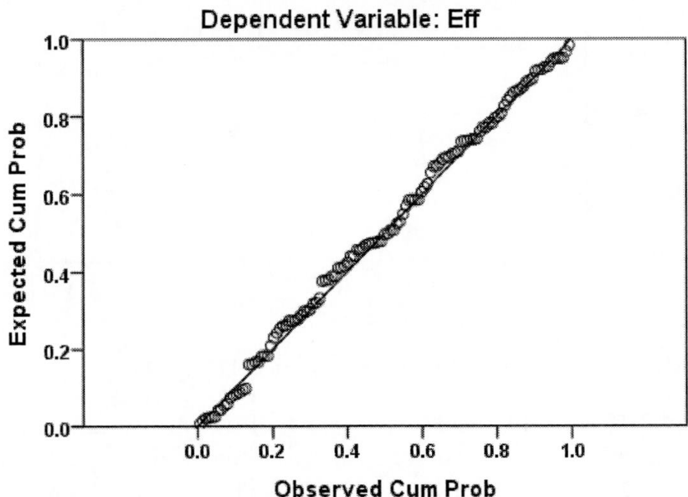

Figure 3.16. Normal probability plot of the regression standardized residuals from linear regression model (3.4) or (3.5).

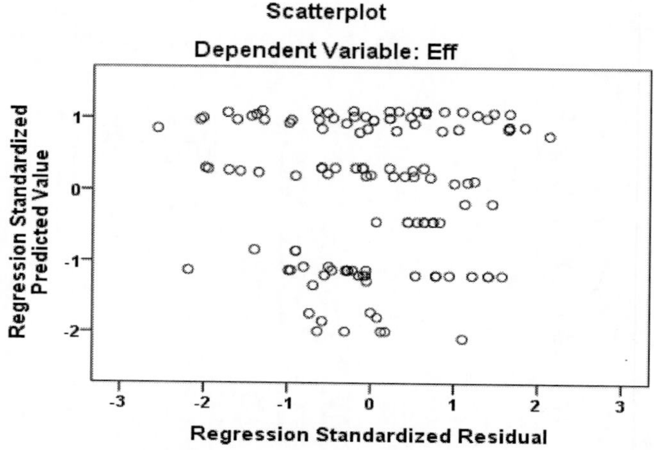

Figure 3.17. Investigation of the homoscedasticity in linear regression model (3.5).

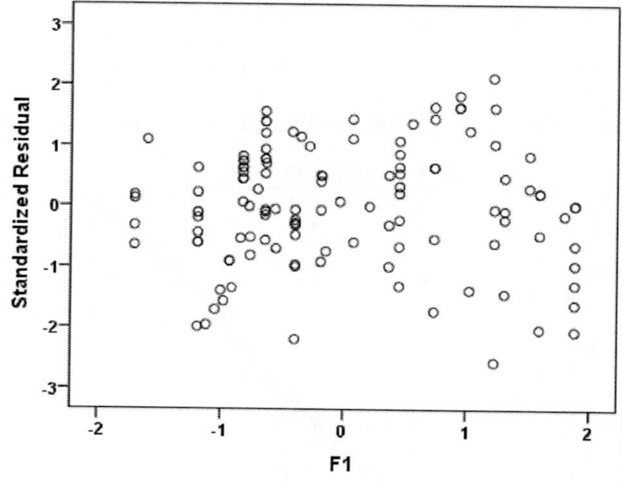

Figure 3.18. Standardized residuals versus *F*1 in model (3.4)-(3.5).

Figure 3.18 – 3.20 display the plots of factor variables against the standardized residuals. The lack of strong patterns in Figure 3.18-3.19 means that *F*1 and *F*2 are not responsible for the possibly violated homoscedasticity requirement. Some weak pattern as a diamond in Figure 3.20 shows the contribution of *F*3={*PH2*} to this problem. As a hole, we may conclude that basic assumptions for conducting regression analysis are more or less satisfied.

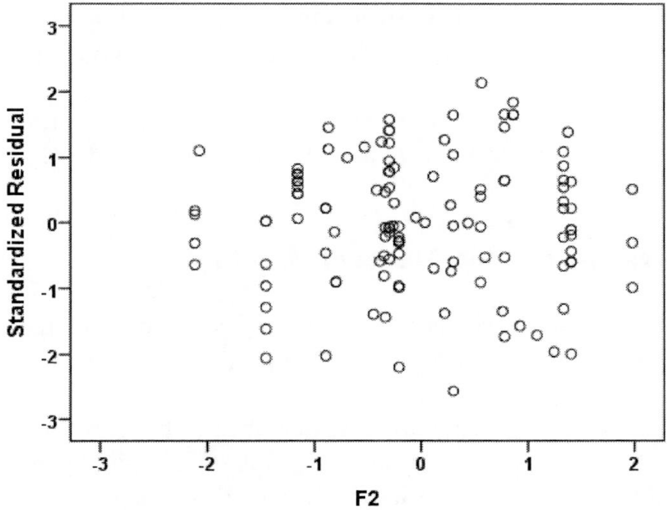

Figure 3.19. Standardized residuals versus *F2* in model (3.4)-(3.5).

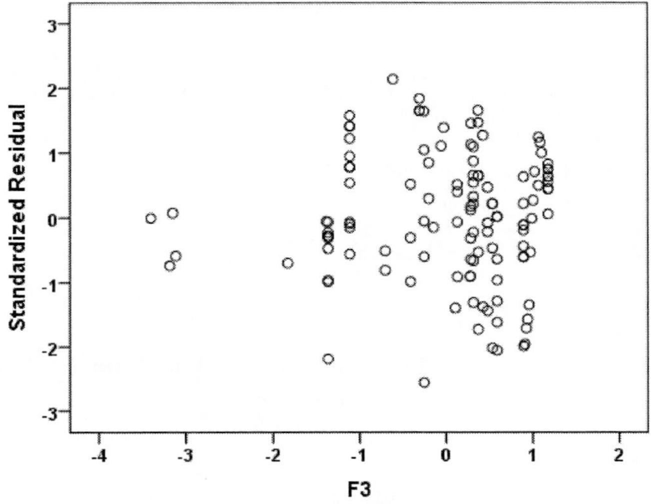

Figure 3.20. Standardized residuals versus *F*3 in model (3.4)-(3.5).

The conducted diagnostic indicates that regression model (3.4)-(3.5) describes correctly the dependence of laser efficiency on the factor variables *F*1, *F*2, *F*3. It needs to be noted that theoretically, when using the Principal Component Regression method, the regression models are biased to some degree [11, 12]. This in addition to the inherent error of experimental measurements of laser characteristics is the main reason for the insufficiently satisfactory predictive accuracy of the model and its relatively high error of over 10%.

Despite the abovementioned weaknesses, the constructed linear model can be regarded as a reference point for the statistical modeling of a CuBr laser.

3.4.2. Linear Regression Model of Output Power

In 3.1 we showed that the logarithmic transformation of the dependent variable *Pout* improves its distribution. The resulting variable was denoted as *LPout* or *Log_Pout.*

In accordance with the procedures, described in the previous section, we will attempt to construct a linear regression model of *LPout* using the predictors *F*1, *F*2, *F*3. The basis indices of the model are given in Tables 3.20 and 3.21. It can be concluded that the linear model is statistically significant and has a coefficient of determination Rsquare=0.887.

Regression equation using coefficients from Table 3.22 is

$$\widehat{LPout} = 3.372 + 0.811 F_1 + 0.285 F_2 + 0.546 F_3 \quad (3.6)$$

For the standardized values, equation is:

$$\widehat{zLPout} = 0.750 F_1 + 0.264 F_2 + 0.505 F_3 \quad (3.7)$$

Table 3.20. Summary of the linear regression model for *LPout* of a CuBr laser

Model Summary[b]				
Model	R	R Square	Adjusted R Square	Std. Error of the Estimate
1	0.942[a]	0.887	0.885	0.36740

[a] Predictors: (Constant), *F*3, *F*2, *F*1.

[b] Dependent Variable: *LPout*.

Table 3.21. ANOVA of the linear regression model for *LPout* of a CuBr laser

ANOVA[b]						
Model		Sum of Squares	df	Mean Square	F	Sig.
1	Regression	124.556	3	41.519	307.588	0.000[a]
	Residual	15.793	117	0.135		
	Total	140.349	120			

[a] Predictors: (Constant), *F*3, *F*2, *F*1.
[b] Dependent Variable: *LPout*.

Table 3.22. Coefficients of the linear regression model for *LPout* of a CuBr laser

Coefficients[a]						
		Unstandardized Coefficients		Standardized Coefficients		
Model		B	Std. Error	Beta	t	Sig.
1	(Constant)	3.372	0.033		100.968	0.000
	F1	0.811	0.034	0.750	24.189	0.000
	F2	0.285	0.034	0.264	8.502	0.000
	F3	0.546	0.034	0.505	16.290	0.000

[a] Dependent Variable: *LPout*.

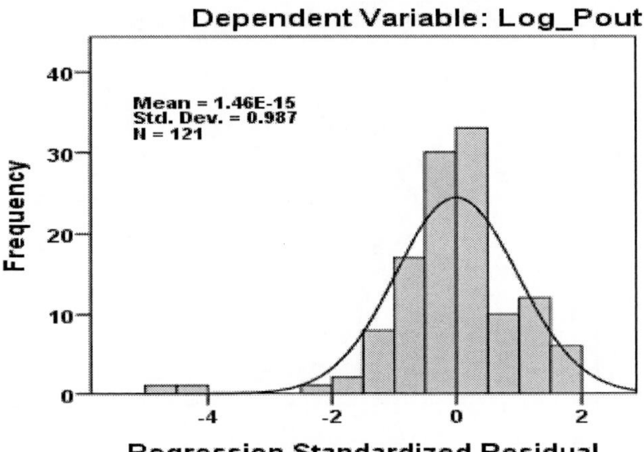

Figure 3.21. Distribution of the residuals in model (3.6)-(3.7).

The investigation of the residuals yielded more unsatisfactory results than those from the previous section. Figure 3.21 shows that the residuals are normally distributed. Figure 3.22 displays the 'Normal probability plot', where the deviations are relatively big, but without heavy tails. Therefore, the residuals are normally distributed. The homoscedasticity condition is more or less violated as shown in Figure 3.23.

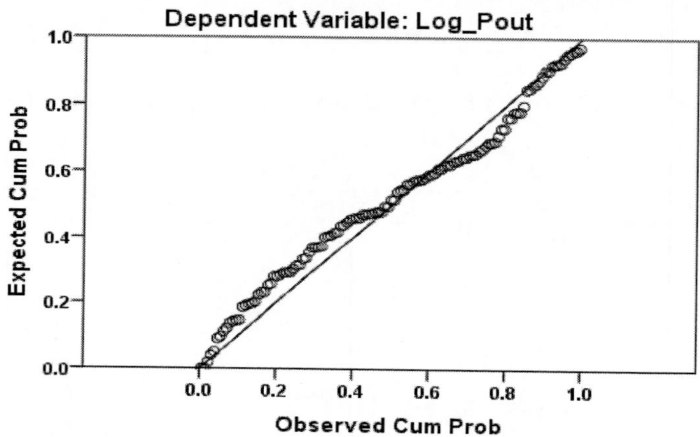

Figure 3.22. Dispersion of expected cumulative probability in model (3.6)-(3.7).

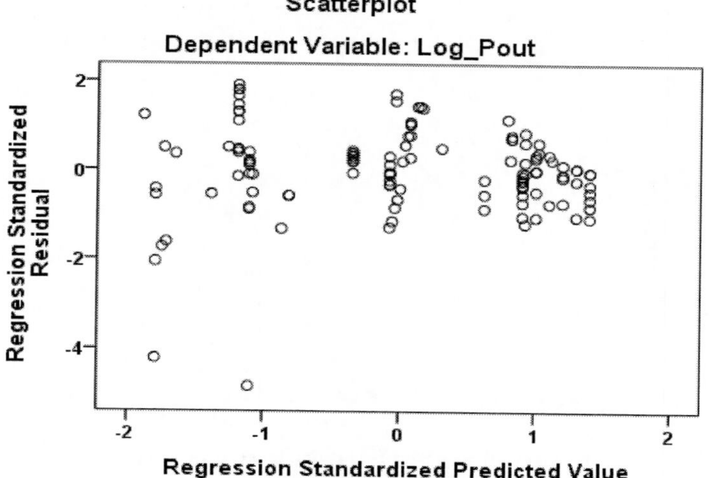

Figure 3.23. Investigation of homoscedasticity in model (3.6)-(3.7).

As a whole, the regression model of *LPout* is inferior in its qualities as compared to the model of *Eff* and results in an error of over 20% when inverse conversion into *Pout* is performed. For these reasons, the linear regression model (3.6)-(3.7) can be considered unsufficiently good. In the next two chapters, we will apply other statistical techniques in order to model the considered data more adequately.

3.5. Physics Interpretation of the Results from Factor and Linear Regression Analyses

The physics interpretation and the comparison between the model and the actual object or phenomenon under investigation are the final validity criterion for every model. The results from factor and regression analyses, confirming the validity of the model are discussed in [2-5, 8-10]. We are going to look into this discussion for above models in more detail.

From Table 3.13 it is seen that the factor *F*1 accounts for 49.5% of the total variance in sample data and therefore is the dominant factor. This combines geometric dimensions *D*, *DR*, *L* and input electric power *PIN*.

The strong influence of the rings (variable *DR*) is explained by the specific role they play in the processes of laser radiation and formation of total laser generation and efficiency. The presence of rings in the tube increases the inner surface area of the tube and improves its cooling and heat balance. The rings limit the active zone of the discharge and enable the formation of a buffer volume between them and the wall of the tube. The high electric pulse concentrates the discharge around the longitudinal axis of the tube. The buffer volume helps to accelerate the processes of plasma relaxation, i.e. the diffusion of charged and neutral particles and their subsequent recombination in the volume or to the walls of the tube. Another very important process takes place in the buffer volume – the intensive recovery of CuBr after the dispersion.

The length of the tube *L* is another geometric component of the model, included in the first factor *F*1. The increase of *L* in the investigated limits influences directly the gain as per the law □ $e^{\alpha L}$. The length and the diameter of the tube *D* influence the input volume power density, and therefore the temperature profile of the gas in the laser tube. What is more, an increase of the inner diameter of the tube *D* increases the surface area and therefore improves the heat balance of the laser tube and decreases the thermal population of lower laser levels. The increase of *D* leads to an increase in the diffusion and drift coefficients of the particles of the tube wall. In this way, the repopulation of the lower laser level is improved. All these processes enhance laser efficiency and laser power.

The fourth component of *F*1 is the input electric power *PIN*, supplied into discharge. Higher values of *PIN* lead to an increase of electron energy and better excitation of upper laser levels. It is an indisputable physical fact that this parameter contributes the most to the enhancement of laser efficiency and power. In total, these four variables participate in the obtained models completely adequately.

The second factor *F*2 has a significantly lower relative share (around 23.65%) of the total variance of the sample (see Table 3.13). Here, the only strongly dominating variable is *PL* – electric power per unit length. Regardless of the fact that *PL* is not a real physical quantity, its participation in the model cannot be overlooked since it is statistically significant and its omission leads to a lower quality of the models. As noted, the influence of *PL* is negative. This variable is energetic in nature. Physically, higher *PL* leads to higher gas temperature which involves the high population of the lower laser states. As results, the laser efficiency drops. In other words, *PL* reflects the fact that the reduction of the radial distribution of the gas temperature increases the efficiency since this distribution of the temperature of the gas is proportional to *PL*. The reason for this is that the radial gradient of the temperature of the gas generates inhomogeneity in the active laser medium, which has an adverse effect on laser operation. By reducing *PL*, the applied electric power in the active medium is increased at the expense of the electrodes. This increases output laser power and efficiency.

As we have seen, the third factor *F*3 is also formed by a single variable (*PH*2 - hydrogen pressure). The positive effect of hydrogen on laser generation and efficiency is long established [13, 14]. The inclusion of this variable in the model also provides an adequate description of actual physical phenomena.

The variables, which correlations with both dependent and other independent variables are statistically unsignificant and for this reason are not included in the constructed models, are: *PRF*, *PNE*, *C* and *TR*. In future studies, as well as during experiments, it is considered that they should be maintained at the established optimal values (this is especially true for *TR* – the temperature of the copper bromide reservoir). The obtained models are valid under these conditions.

We will summarize the results from the FA and RA based on experimental data. The newly established quantitative interactions between laser efficiency (or output power) and the six basic laser variables *PIN*, *DR*, *L*, *D*, *PL* and *PH2* are as follows:

- The four variables *PIN*, *DR*, *L* and *D* from the first factor need to be regarded as a group where their priority and relative influence on *Eff* and *LPout* (respectively on *Pout*) is approximately described by their factor loadings from the rotation matrix in Table (3.16).
- The total relative influence of grouped variables according to factors on laser efficiency is approximately described by the coefficients in the linear regression model (3.4).

Despite the relatively not very high accuracy of the approximation, the obtained models are acceptable and can be considered valid as essentially they describe the experimental data. The actual physical processes related to the increase of laser efficiency are adequately modeled.

As far as the use of the models for experiment planning, the following conclusions can be drawn. The increase of laser efficiency can be achieved by simultaneously increasing the geometric dimensions and input electric power within the established approximate proportions. With regard to hydrogen pressure, the addition of which element in small quantities affects output laser characteristics significantly, there is room for further experimental investigation in order to determine the optimal value of this laser characteristic at different geometric and energetic conditions.

In general, with the help of parametric linear models, the obtained results indicated a reduction in the dimensionality of the predominant part of the initial data from 12 to 6 variables. The degree of influence of each basis variable on the value of laser efficiency and power was determined. Regardless of unsufficient accuracy of some of the constructed linear models, this type of estimates of output laser characteristics is fully useful as initial description of the main dependences. These type of qualitative results derived from experiments cannot be obtained through kinetic and other structural methods, mentioned in Chapter 1.

REFERENCES

[1] http://www.spss.com/software/statistics/stats-pro/, IBM SPSS Statistics, 2011.

[2] P. Iliev, S. G. Gocheva-Ilieva and N. V. Sabotinov, *Classification analysis of the variables in a CuBr laser, Quantum Electron.* 39(2) (2009) 143-146.

[3] P. Iliev, S. G. Gocheva-Ilieva, D. N. Astadjov, N. P. Denev and N. V. Sabotinov, *Statistical approach in planning experiments with a copper bromide vapor laser, Quantum Electron.* 38(5) (2008) 436-440.

[4] P. Iliev, S. G. Gocheva-Ilieva, D. N. Astadjov, N. P. Denev and N. V. Sabotinov, *Statistical analysis of the CuBr laser efficiency improvement, Opt. Laser Technol.* 40(4) (2008) 641-646.

[5] S. G. Gocheva-Ilieva and I. P. Iliev, *Parametric and nonparametric empirical regression models of copper bromide laser generation, Math. Probl. Eng., Theory, Methods and Applications,* Hindawi Publ. Corp., New York, NY, Article ID 697687 (2010), 15 pages.

[6] T. P. Ryan, *Modern engineering statistics*, John Wiley and Sons, Inc., Hoboken, New Jersey, 2007.

[7] C. Rencher, *Methods of multivariate analysis*, 2 ed., John Wiley, New York, 2002.

[8] .P. Iliev, S. G. Gocheva-Ilieva, N. P. Denev and N. V. Sabotinov, *Statistical Study of the Copper Bromide Laser Efficiency*, Proc. of Sixth Intern. Conf. of the Balkan Physical Union, Istanbul, Turkey, in: CP899, Melville NY: American Institute of Physics, p. 680, 2007.

[9] P. Iliev and S. G. Gocheva-Ilieva, *Statistical techniques for examining copper bromide laser parameters*, Proc. of Int. Conf. of Numerical Analysis and Applied Mathematics, ICNAAM 2007, Corfu, Greece, In: CP936, Melville NY: American Institute of Physics, pp. 267-270, 2007.

[10] P. Iliev and S. G. Gocheva-Ilieva, *On the application of the multidimensional statistical techniques for exploring copper bromide vapor laser*, Proc. of the 34th Conf. on Applications of Mathematics in Engineering and Economics (AMEE'08), Sozopol, Bulgaria, in: CP1067, Melville NY: American Institute of Physics, pp. 475-482, 2008.

[11] T. Hastie, R. Tibshirani and J. Friedman, *The Elements of Statistical Learning: Data Mining, Inference and Prediction*, 2nd ed., Springer, New York, 2009.

[12] J. Izenman, *Modern multivariate statistical techniques: regression, classification, and manifold learning*, Springer, New York, 2008.

[13] D. N. Astadjov, N. V. Sabotinov, N. K. Vuchkov, *Effect of hydrogen on CuBr laser power and efficiency*, *Opt. Commun.* 56(4) (1985) 279–282.

[14] N. V. Sabotinov, *Metal vapor lasers*, in: *Gas Lasers*, eds. M. Endo, R. F. Walter, CRC Press, Boca Raton, 2006.

Chapter 4

POLYNOMIAL AND NONLINEAR MODELS OF THE OUTPUT CHARACTERISTICS OF A COPPER BROMIDE VAPOR LASER

ABSTRACT

The goal of this chapter is to obtain parametric regression models better than the linear models presented in Chapter 3.

The models are constructed for the same model sample as the one in Chapter 3 using derived factor variables and the laser variables, which have not been included so far. The influence of factor variables up to the third degree for the polynomial models is accounted for.

The obtained parametric models are compared. The results indicate that nonlinear models demonstrate better statistical indices than all types of parametric models.

Some preliminary results from another sample of similar quality are published in [1, 2].

The models are obtained with the help of SPSS and the Wolfram *Mathematica* software [3, 4].

4.1. RELATIONSHIP BETWEEN FACTOR VARIABLES AND OUTPUT CHARACTERISTICS

The raw trends in the behavior of *Eff* and *Pout* dependent on factors F_1, F_2, F_3 are linear and the models in Chapter 3 were constructed on this basis. Of course, in more detail, the dependences contain nonlinearities. First, we will attempt to investigate these graphically. Later on, we will describe the main relationships using the capabilities of parametric modeling.

4.1.1. Relationship between Factor Variables and Efficiency

The three factors F_1, F_2, F_3 affect laser efficiency *Eff* differently. Figure 4.1 a), b), c) illustrates the relationships between each of the factors and *Eff*, as well as the LOESS smoothing curves, obtained using the Epanechnikov kernel for 50% of the points.

a)

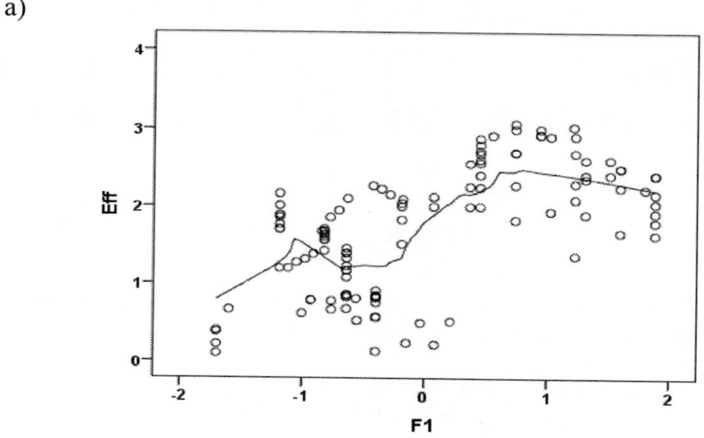

b)

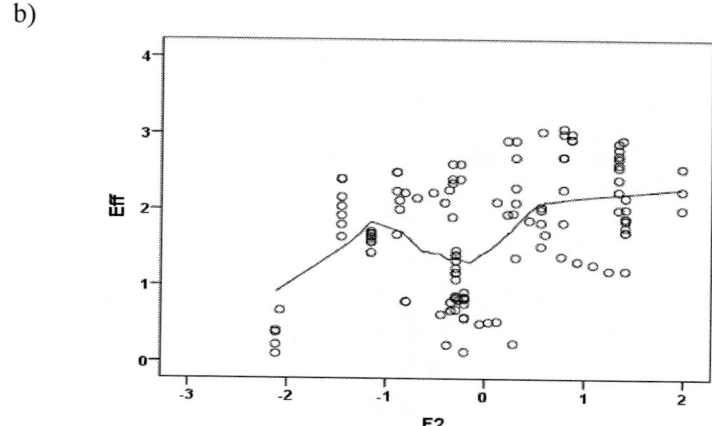

Figure 4.1. (Continued).

c)

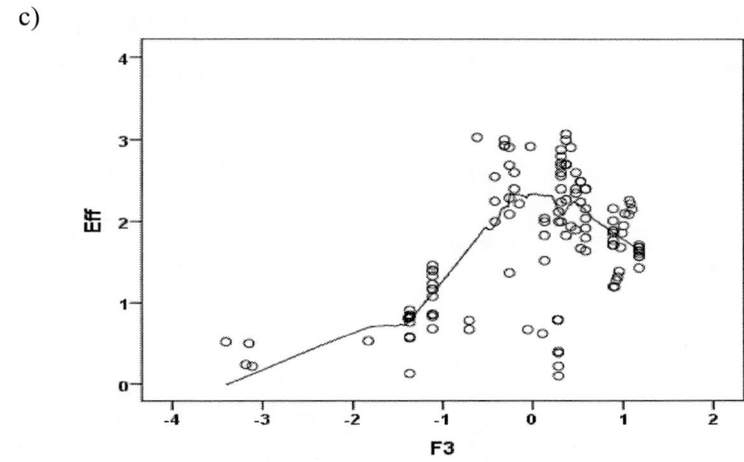

Figure 4.1. a)-c). Common relationship between *Eff* and each of the three PCA factors F_1, F_2, F_3 and corresponding LOESS smoothing Epanechnikov curves.

Figure 4.1 a) shows that as a whole the general relationship between *Eff* and the dominant factor F_1 is linear with increasing trend. The same can also be concluded for the second factor. The relationship with the third factor in the right part of the graph is a descending one, which follows from the greater part of the data. The more detailed smoothing curves contain elements of second and third order polynomials. We can conclude that in addition to the linear terms we should also be taking into account second- and even third order interactions between factor variables when constructing polynomial or nonlinear regression models.

4.1.2. Relationship between Factor Variables and Output Power

Factor variables F_1, F_2, F_3 affect laser output power *Pout* in a way similar to the one described above with regard to laser efficiency. The general behavior approximates closely the presented two- and three-dimensional plots of the dependences within previous models [2].

The type of relationships between the logarithm of *Pout* and the factor variables are shown in Figure 4.2 a), b), c).

In short, the raw trends for F_1 and F_2 are ascending linear ones, with nonlinearities, and in the case of F_3, the detailed relationship are polynomial.

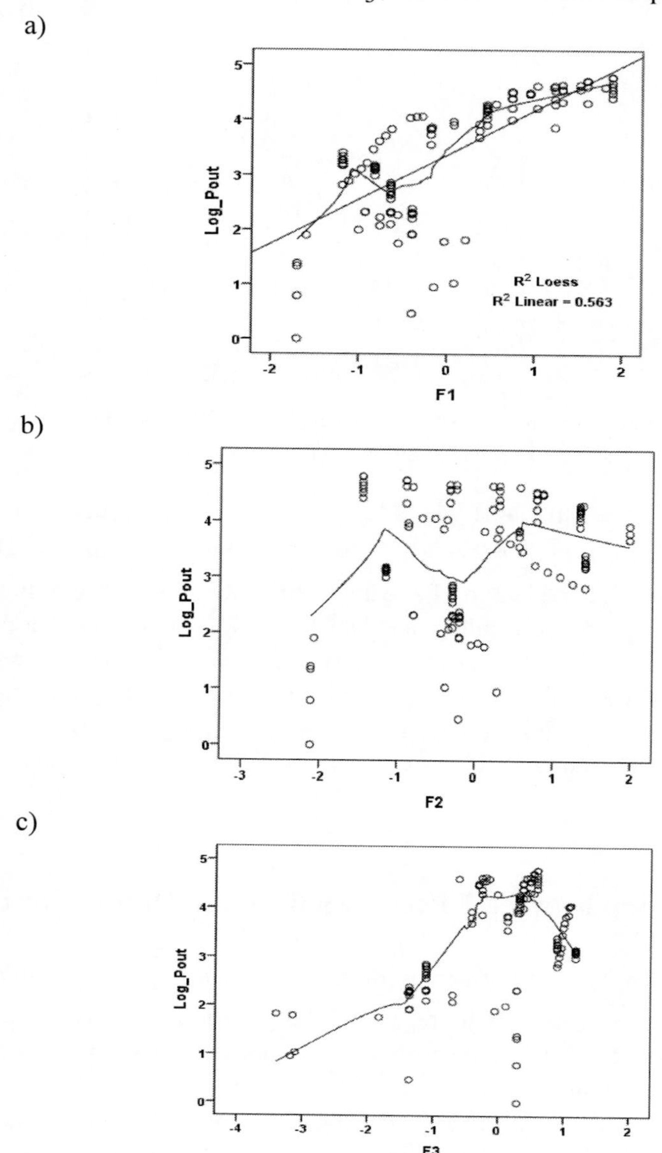

Figure 4.2. a)-c). Common relationship between *LPout* and each of the three PCA factors F_1, F_2, F_3 and corresponding LOESS smoothing Epanechnikov curves.

4.2. Polynomial Type Regression Models

In this section, we will attempt to improve the models presented in Chapter 3 by using the factors and their first and second order products, as well as to add laser variables, which have not been included up until now in the analysis for the model construction.

Based on general conclusions in 4.1, in an exploratory manner, we will construct polynomial regression models by consecutively using the following three groups of predictors:

The first group of predictors is consisting of:

$$F_1, F_2, F_3, PRF, PNE, C, TR. \tag{4.1}$$

The second group of predictors accounts for interactions between factors up to the second degree and additional variables:

$$F_1,\ F_2,\ F_3,\ F_1^2,\ F_2^2,\ F_3^2,\ F_1F_2,\ F_1F_3, F_2F_3,\ PRF, PNE, C, TR. \tag{4.2}$$

The third group of predictors accounts for influences up to the third degree of the factor variables and additional variables:

$$\begin{gathered} F_1,\ F_2,\ F_3,\ F_1^2,\ F_2^2,\ F_3^2,\ F_1F_2,\ F_1F_3,\ F_2F_3,\ F_1^3,\ F_2^3,\ F_3^3,\ F_1^2F_2,\ F_1^2F_3 \\ F_2^2F_3,\ F_2^2F_1, F_3^2F_1,\ F_3^2F_2,\ F_1F_2F_3,\ PRF, PNE, C, TR \end{gathered} \tag{4.3}$$

In order to construct the model, we applied classical linear regression methodology, as in Chapter 3. The stepwise regression method was used to exclude the statistically insignificant predictors. The detailed calculations performed in order to obtain the results are similar to those in the previous chapter and have been omitted.

We have to note that the inclusion of interactions between the factors and the additional variables does not alter the models. To this reason, these models have not been described with regard to laser efficiency as well as to output power.

4.2.1. Polynomial Regression Models of Laser Efficiency

The following polynomial models of the efficiency *Eff* were obtained, respectively for the groups of predictors (4.1) – (4.3):

$$\widehat{Eff} = 1.717 + 0.484\,F_1 + 0.316\,F_2 + 0.385\,F_3 \tag{4.4}$$

$$\widehat{Eff} = 1.833 + 0.574\,F_1 + 0.279\,F_2 + 0.441F_3 + 0.096F_1F_2 - 0.117F_1^2 \tag{4.5}$$

$$\widehat{Eff} = 1.8819 + 0.510\,F_1 + 0.267\,F_2 + 0.552\,F_3 - 0.309F_1^2F_3 \tag{4.6}$$

The basic statistical indices of these models are given in Table 4.1. The corresponding models in standardized form are

$$\widehat{zEff} = 0.618\,F_1 + 0.404F_2 + 0.491F_3 \tag{4.7}$$

$$\widehat{zEff} = 0.734\,F_1 + 0.359\,F_2 + 0.563\,F_3 + 0.149F_1F_2 - 0.151F_1^2 \tag{4.8}$$

$$\widehat{zEff} = 0.652\,F_1 + 0.341F_2 + 0.667\,F_3 - 0.284F_1^2F_3 \tag{4.9}$$

Here *zEff* are the standardized variables of *Eff*.

Table 4.1. Summary of the polynomial models of the laser efficiency *Eff* of a CuBr laser

Model number	Group of predictors	*R*2	*R*2 adj.	Stand. Error of the Estimate	Predictors in the model
(4.4)	(4.1)	0.786	0.781	0.36655	F_1, F_2, F_3
(4.5)	(4.2)	0.825	0.818	0.33440	$F_1, F_2, F_3, F_1F_2, F_1^2$
(4.6)	(4.3)	0.831	0.825	0.32741	$F_1, F_2, F_3, F_1^2F_3$

It is obvious that model (4.4) coincides with model (3.3). None of the models contain the additional variables *PRF*, *PNE*, *C*, *TR*, since they were found to be statistically insignificant.

4.2.2. Polynomial Regression Models of *LPout*

In the previous chapter we showed that the considered sample as well as all available data of output laser power are not normally distributed. For this reason, formally we cannot apply polynomial regression for *Pout*. Analogically to Chapter 3, here also, we will use the logarithmic transformation *LPout* of *Pout* (also denoted by *Log_Pout*), the distribution of which is closed to normal.

By using the three groups of predictors (4.1) – (4.3) and applying stepwise regression method, we obtain the following regression models, respectively:

$$\widehat{LPout} = 3.787 + 0.784\,F_1 + 0.241\,F_2 + 0.587\,F_3 - 0.021PNE \qquad (4.10)$$

$$\begin{aligned}\widehat{LPout} = {} & 3.832 + 0.832\,F_1 + 0.195\,F_2 + 0.619\,F_3 \\ & - 0.018PNE - 0.113F_1^2\end{aligned} \qquad (4.11)$$

$$\begin{aligned}\widehat{LPout} = {} & 3.850 + 0.856\,F_1 + 0.955\,F_3 - 0.014PNE - 0.590F_1^2 F_3 \\ & - 0.381F_1F_2F_3 - 0.024F_3^3 + 0.059F_2^2 - 0.162F_1^2\end{aligned} \qquad (4.12)$$

The corresponding equations in standardized form are:

$$\widehat{zLPout} = 0.725\,F_1 + 0.223\,F_2 + 0.543\,F_3 - 0.167\,zPNE \qquad (4.13)$$

$$\widehat{zLPout} = 0.769\,F_1 + 0.180\,F_2 + 0.572\,F_3 - 0.140\,zPNE - 0.105F_1^2 \qquad (4.14)$$

$$\begin{aligned}\widehat{zLPout} = {} & 0.791\,F_1 + 0.883\,F_3 - 0.111\,zPNE - 0.392F_1^2 F_3 \\ & - 0.248F_1F_2F_3 - 0.136F_3^3 + 0.063F_2^2 - 0.151F_1^2\end{aligned} \qquad (4.15)$$

Here *zLPout* and *zPNE* are standardized variables of *LPout* and *PNE*, respectively.

The summaries of constructed polynomial models (4.10)-(4.12) are given in Table 4.2. All obtained models exhibit statistical significance Sig.=0 and have been validated in the same manner as in Chapter 3. Of the additional variables, only PNE was found to be statistically significant. However, as indicated by models (4.13)-(4.15), its relative weight in the models is relatively small. The basic dominating variables in descending order according to their weight are: F_1, F_3 and F_2. This corresponds entirely with the general picture of the laser characteristics grouped in those factors and the respective experimental data.

With the exception of 4-5 outliers, the accuracy of model (4.10) is satisfactory and is within 10-12%, calculated for laser output power *Pout*. It needs to be noted that despite the higher values of Rsquare and Rsqure adjusted, model (4.11) and in particular model (4.12), demonstrate indices that are more unstable when the residuals and the heteroscedsticity are diagnosed.

Table 4.2. Summary of the polynomial models of the variable *LPout* for a CuBr laser

Model number	Group of predictors	*R2*	*R2* adj.	Stand. Error of the Estimate	Predictors in the model
(4.10)	(4.1)	0.912	0.909	0.32701	F_1, F_2, F_3, PNE
(4.11)	(4.2)	0.917	0.913	0.31889	$F_1, F_2, F_3, F_1^2, PNE$
(4.12)	(4.3)	0.936	0.932	0.28229	$F_1, F_3,\ F_1^2 F_3,\ F_1 F_2 F_3,\ F_3^3,\ F_2^2,\ F_1^2,\ PNE$

We have to add that complicating the model further with higher order members, yields worse predicting ability. At this stage, we can be satisfied of the first or second degree polynomial models (4.10) and (4.11) which offer the best approximations of *Pout*. These models show better diagnostics as regard to the residual diagnostics than model (4.12).

4.3. Parametric Nonlinear Regression Models

As mentioned earlier, the models constructed so far, using linear and polynomial type regression with principal components can be considered as the first and second approximations describing the relationships of interest. For this reason, we set the task of find nonlinear regression models of laser efficiency and output power.

Various transformations are used in statistics in order to improve the multivariate distribution of data [5, 6]. In our case, the standardized factor loadings have both positive and negative values. For this reason, they are transformed using the Yeo-Johnson transformation (see (2.11) in Chapter 2).

The nonlinear model we are looking for is in the quasilinear common form

$$\hat{Y}(\theta,\lambda)=\theta_0+\theta_1\psi_{YJ}(\lambda_1,X_1)+...+\theta_p\psi_{YJ}(\lambda_p,X_p) \qquad (4.16)$$

where $\hat{Y}$ is the dependent variable, $X_1,...,X_p$ are the independent data variables, ψ_{YJ} is the Yeo-Johnson transformation and $\theta_0,\theta_1,...,\theta_p;\ \lambda_1,\lambda_2,...,\lambda_p$ are the unknown parameters of the model. The parameters have to be estimated using the least squares method.

We will also construct the nonlinear models, which can contain second or third degree terms of Yeo-Johnson transformations, such as $(\psi_{YJ}(\lambda_i,X_i))^2$ and $(\psi_{YJ}(\lambda_i,X_i))^3$.

It must be noted, that the general assumption with nonlinear model fit is that the original dependent variable is normally distributed. For this reason, further on, we will once again model *LPout* instead of *Pout.*

The nonlinear models are constructed with the help of the three factor variables F_1,F_2,F_3. All calculations are carried out by Wolfram *Mathematica* 7 software [2].

4.3.1. Nonlinear Regression Models of Laser Efficiency

We will consider three nonlinear models of laser efficiency *Eff.*

The first nonlinear model for estimation of *Eff* is represented in the form, containing linearly F_1 and transformed F_2,F_3:

$$\widehat{Eff}(\theta,\lambda)=\theta_0+\theta_1F_1+\theta_2\psi_{YJ}(\lambda_2,F_2)+\theta_3\psi_{YJ}(\lambda_3,F_3) \qquad (4.17)$$

where $\theta_0,\theta_1,\theta_2,\theta_3;\lambda_2,\lambda_3$ are the unknown parameters. To determine them the compact code in Wolfram *Mathematica* was written (see Figure 4.3).

```
(*  Yeo-Johnson transformation: *)

In[1]:= ψ[λ_, y_] := If[y ≥ 0 && λ ≠ 0, ((y + 1)^λ - 1)/λ,
      If[y < 0 && λ ≠ 2, -((1 - y)^(2-λ) - 1)/(2 - λ),
       If[y ≥ 0 && λ == 0, Log[y + 1],
        If[y < 0 && λ == 2, -Log[1 - y]]]]]
n = 121;
f1 = ReadList["f1.txt", Number] ;
f2 = ReadList["f2.txt", Number] ;
f3 = ReadList["f3.txt", Number] ;
pout = ReadList["pout.txt", Number] ;
eff = ReadList["eff.txt", Number] ;
data = Table[{f1[[i]], f2[[i]], f3[[i]], eff[[i]]},
   {i, 1, n}];
nlmodel = NonlinearModelFit[data, theta0 + theta1 * x1 +
   theta2 * ψ[λ2, x2] + theta3 * ψ[λ3, x3],
  {theta0, theta1 , theta2, theta3, λ2, λ3}, {x1, x2, x3 }];
nlmodel["BestFitParameters"];
predicted = nlmodel["PredictedResponse"];
nlmodel["ANOVATable"]
nlmodel["ParameterTable"]
nlmodel["FitCurvatureTable"]
Export["C:\\Users\\Snow\\Eff_nlm1\\predEff-m1.txt",
predicted, "List"]
```

Figure 4.3. *Mathematica* code for the nonlinear model (4.17) of laser efficiency of CuBr lasers.

The calculations were performed with double precision accuracy on a dual core personal computer and took around 5 seconds to compute. The results in their original form, obtained using *Mathematica* are shown in Figure 4.4.

All information about the model is stored in a variable *nlmodel* (see Figure 4.3 and corresponding outputs in Figure 4.4). The first line returns a symbolic 'FittedModel' object to represent the constructed model. With the help of the 'PredictedResponse' property, we obtain the predicted values of *Eff*, in this case they are assigned to the variable *predicted* and are not visualized.

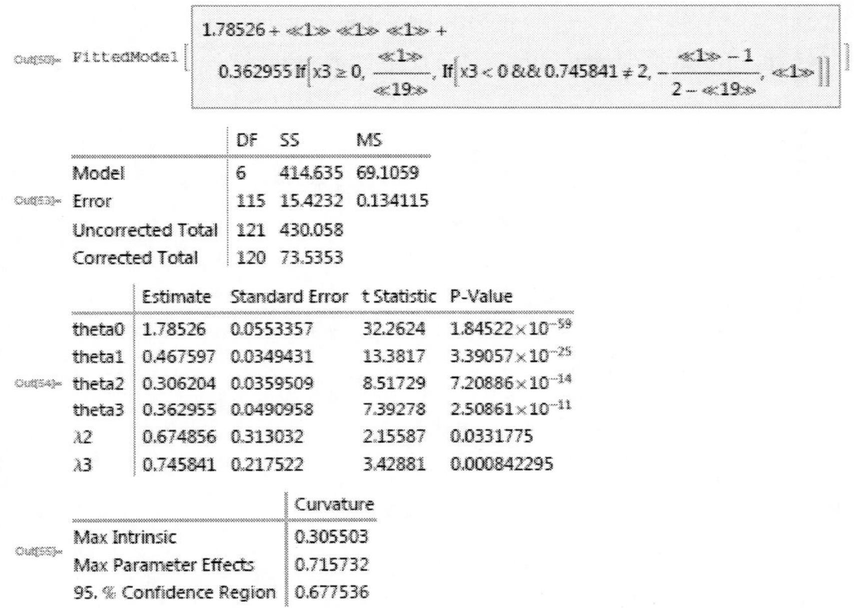

Out[50]= FittedModel[1.78526 + «1» «1» «1» + 0.362955 If[x3 ≥ 0, «1» / «19», If[x3 < 0 && 0.745841 ≠ 2, − («1» − 1) / (2 − «19»), «1»]]]

Out[53]=

	DF	SS	MS
Model	6	414.635	69.1059
Error	115	15.4232	0.134115
Uncorrected Total	121	430.058	
Corrected Total	120	73.5353	

Out[54]=

	Estimate	Standard Error	t Statistic	P-Value
theta0	1.78526	0.0553357	32.2624	1.84522×10^{-59}
theta1	0.467597	0.0349431	13.3817	3.39057×10^{-25}
theta2	0.306204	0.0359509	8.51729	7.20886×10^{-14}
theta3	0.362955	0.0490958	7.39278	2.50861×10^{-11}
λ2	0.674856	0.313032	2.15587	0.0331775
λ3	0.745841	0.217522	3.42881	0.000842295

Out[55]=

	Curvature
Max Intrinsic	0.305503
Max Parameter Effects	0.715732
95. % Confidence Region	0.677536

Figure 4.4. Output results of the calculation of the nonlinear regression model (4.17) with the help of the Wolfram *Mathematica* software.

From the second output in Figure 4.4 a brief report of the resulting ANOVA table is given. The next output is the 'ParameterTable'. The first column 'Estimate' contains the coefficients of the nonlinear six-dimensional model (4.17). The last column of this table shows that all coefficients are statistically significant (P-value<0.05) and therefore the model is correct. The last output of Figure 4.4 contains information about the curvature of the model. The value of the parameter-effects could be above the critical value of '95 % Confidence Region'. This indicates that the least-squares parameters describe a nonlinear behavior, rather than linear.

With the help of the build-in function NonlinearModelFit[] of *Mathematica,* a total of about 50 properties related to data and the fitted function can be calculated [7].

The second nonlinear model of *Eff* is suggested by the second degree polynomial model (4.5). We reparametrized the model in the following form

$$\widehat{Eff}(\theta,\lambda)=\theta_0+\theta_1\psi_{YJ}(\lambda_1,F_1)+\theta_2\psi_{YJ}(\lambda_2,F_2)+\theta_3\psi_{YJ}(\lambda_3,F_3)+\theta_4\left(\psi_{YJ}(\lambda_1,F_1)\right)^2 \qquad (4.18)$$

where $\theta_0, \theta_1, \theta_2, \theta_3, \theta_4; \lambda_1, \lambda_2, \lambda_3$ are the new unknown parameters.

We modify the code and obtain the results in Figure 4.5.

Out[76]= 0.114158

Out[77]=

	Estimate	Standard Error	t Statistic	P-Value
theta0	1.88911	0.0666795	28.3312	3.66065×10^{-53}
theta1	0.71537	0.0681635	10.4949	2.1278×10^{-18}
theta2	0.192728	0.0425335	4.53121	0.0000146429
theta3	0.407864	0.045583	8.94771	8.27038×10^{-15}
theta4	−0.240028	0.0405141	−5.92455	3.45277×10^{-8}
λ1	1.55249	0.181725	8.54311	7.00441×10^{-14}
λ2	1.23609	0.549594	2.24909	0.026445
λ3	0.831591	0.190142	4.37352	0.0000273443

Out[78]=

	Curvature
Max Intrinsic	0.65168
Max Parameter Effects	1.93057
95. % Confidence Region	0.703364

Figure 4.5. Output results of the nonlinear regression model (4.18) with the help of the Wolfram *Mathematica* software.

The first row contains an estimate of the error variance. In the next output we can see the parameter table. The values of the model are given in the first column. The last – fourth column - indicates that all of these are statistically significant at the 0.05 level. The last output line shows that 'Max Parameter Effects' is almost 2.5 times higher than the '95% Confidence Region", which means the model essentially accounts for the nonlinearities of the data of laser efficiency *Eff*.

However, using all predictors from model (4.5) did not produce a satisfactory result and is not presented here.

The third nonlinear model of *Eff* is suggested by the polynomial model (4.6) and use the same predictors. The model is

$$\widehat{Eff}(\theta,\lambda) = \theta_0 + \theta_1 \psi_{YJ}(\lambda_1, F_1) + \theta_2 \psi_{YJ}(\lambda_2, F_2) + \theta_3 \psi_{YJ}(\lambda_3, F_3) + \theta_4 \left(\psi_{YJ}(\lambda_1, F_1)\right)^2 \psi_{YJ}(\lambda_3, F_3) \quad (4.19)$$

where $\theta_0, \theta_1, \theta_2, \theta_3, \theta_4; \lambda_1, \lambda_2, \lambda_3$ are the relative unknown parameters.

We modify partially the code and visualize some additional properties of the function NonlinearModelFit. The code is shown in Figure 4.6.

```
nlmodel = NonlinearModelFit[data, theta0 + theta1 * ψ[λ1, x1] +
    theta2 * ψ[λ2, x2] + theta3 * ψ[λ3, x3] +
    theta4 * ψ[λ1, x1] * ψ[λ1, x1] * ψ[λ3, x3],
   {theta0, theta1 , theta2, theta3, theta4, λ1, λ2, λ3},
   {x1, x2, x3 }];
 nlmodel["BestFitParameters"];
 predicted = nlmodel["PredictedResponse"];
 nlmodel["EstimatedVariance"]
 nlmodel["ParameterTable"]
 nlmodel["FitCurvatureTable"]
 nlmodel["CorrelationMatrix"] // MatrixForm
 nlmodel["ParameterBias"] // MatrixForm
 nlmodel["ANOVATable"]
 Export["C:\\Users\\Snow\\Eff_nlm1\\predEff-m3.txt",
    predicted, "List"]
```

Figure 4.6. *Mathematica* code for the nonlinear model (4.19) of laser efficiency of CuBr lasers.

The parameter table of the model (4.19) is shown in Figure 4.7. The results are completely satisfactory.

Out[83]= 0.105006

Out[84]=

	Estimate	Standard Error	t Statistic	P-Value
theta0	1.74666	0.0550848	31.7086	4.46234×10^{-58}
theta1	0.539802	0.039345	13.7197	8.23167×10^{-26}
theta2	0.246594	0.0400323	6.15988	1.14678×10^{-8}
theta3	0.597495	0.0536099	11.1452	6.52862×10^{-20}
theta4	−0.376902	0.0607559	−6.20355	9.32666×10^{-9}
λ1	0.969599	0.122707	7.90174	1.98701×10^{-12}
λ2	1.62582	0.427374	3.80422	0.000231294
λ3	1.34959	0.175647	7.68353	6.11058×10^{-12}

Out[85]=

	Curvature
Max Intrinsic	0.550287
Max Parameter Effects	1.28022
95. % Confidence Region	0.703364

Figure 4.7. Output results of the nonlinear regression model (4.19) with the help of the Wolfram *Mathematica* software.

The outputs from the additional properties are given in Figure 4.8. There is small asymptotic correlation between all parameters, the higher one is 0.702 between θ_2, λ_2. This indicates that the model is appropriate for our data. The last output displays the 'ParameterBias' which gives a vector of the estimated differences between the parameter estimates and the true parameter values. Parameter bias is based on the average curvature of the solution locus tangential to the least-squares estimates.

$$\begin{pmatrix} 1 & -0.403 & 0.275 & -0.032 & -0.046 & -0.384 & -0.513 & -0.542 \\ -0.403 & 1 & -0.200 & 0.147 & -0.189 & 0.528 & 0.200 & 0.386 \\ 0.275 & -0.200 & 1 & -0.167 & 0.429 & -0.085 & -0.702 & -0.153 \\ -0.032 & 0.147 & -0.167 & 1 & -0.637 & -0.214 & 0.174 & 0.507 \\ -0.046 & -0.189 & 0.429 & -0.637 & 1 & 0.084 & -0.461 & -0.231 \\ -0.384 & 0.528 & -0.085 & -0.214 & 0.084 & 1 & -0.075 & 0.083 \\ -0.513 & 0.200 & -0.702 & 0.174 & -0.461 & -0.075 & 1 & 0.283 \\ -0.542 & 0.386 & -0.153 & 0.507 & -0.231 & 0.083 & 0.283 & 1 \end{pmatrix} \begin{pmatrix} -0.00141 \\ -0.00580 \\ -0.00600 \\ -0.00350 \\ 0.00560 \\ 0.00466 \\ 0.02741 \\ -0.00117 \end{pmatrix}$$

Figure 4.8. Correlation matrix and parameter bias of the nonlinear regression model (4.19).

Figure 4.9 shows the plot of *Eff* against its values predicted by model (4.19).

The statistical indices of the constructed nonlinear models of *Eff* are given in Table 4.3.

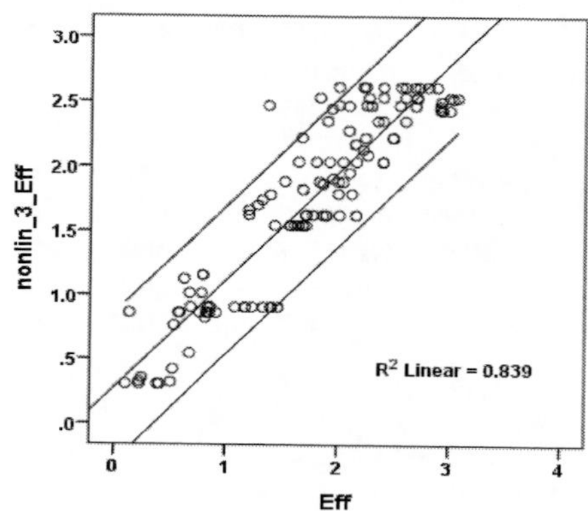

Figure 4.9. Comparison between the experimental values of *Eff* and those calculated using the nonlinear model (4.19).

Table 4.3. Statistics of the constructed nonlinear models of laser efficiency of CuBr lasers

Model number	$R2$	$R2$ adj.	Predictors in the model
(4.17)	0.790	0.788	F_1, F_2, F_3
(4.18)	0.825	0.823	F_1, F_2, F_3, F_1^2
(4.19)	0.839	0.837	$F_1, F_2, F_3, F_1^2 F_3$

4.3.2. Nonlinear Regression Models of Output Power

We will present three nonlinear models of variable *LPout* = ln(*Pout*). Their basic statistics are given in Table 4.4.

The first nonlinear model for estimation of *LPout* is represented in the form

$$\hat{L}Pout(\theta,\lambda) = \theta_0 + \theta_1 F_1 + \theta_2 \psi_{YJ}(\lambda_2, F_2) + \theta_3 \psi_{YJ}(\lambda_3, F_3) \qquad (4.20)$$

where $\theta_0, \theta_1, \theta_2, \theta_3; \lambda_2, \lambda_3$ are the unknown parameters. They are determined as in the case of laser efficiency in model (4.17). Their values are:

$$\theta_0 = 3.54263,\ \theta_1 = 0.765977,\ \theta_2 = 0.180851,\ \theta_3 = 0.577457$$
$$\lambda_2 = -0.631409,\ \lambda_3 = 0.784382$$

It must be noted that coefficient λ_2 has P-value=0.1211 and is not statistically significant.

The second constructed nonlinear model is of the following form, suggested partially by model (4.12):

$$\begin{aligned}\hat{L}Pout(\theta,\lambda) = \theta_0 &+ \theta_1 \psi_{YJ}(\lambda_1, F_1) + \theta_3 \psi_{YJ}(\lambda_3, F_3) \\ &+ \theta_5 \left(\psi_{YJ}(\lambda_1, F_1)\right)^2 \psi_{YJ}(\lambda_3, F_3) \\ &+ \theta_6 \psi_{YJ}(\lambda_1, F_1) \psi_{YJ}(\lambda_2, F_2) \psi_{YJ}(\lambda_3, F_3)\end{aligned} \qquad (4.21)$$

The coefficients of the model and their "ParameterTable' are shown in Figure 4.8:

	Estimate	Standard Error	t Statistic	P-Value
theta0	3.56388	0.0413618	86.1636	6.52129×10^{-105}
theta1	0.88176	0.0393274	22.421	2.07538×10^{-43}
theta3	0.924351	0.0617027	14.9807	1.31293×10^{-28}
theta5	−0.192807	0.0862897	−2.23441	0.0274233
theta6	−0.668685	0.0781086	−8.56097	6.3764×10^{-14}
λ1	0.583594	0.0859588	6.78924	5.49746×10^{-10}
λ2	3.21214	0.769359	4.17508	0.0000588237
λ3	1.29316	0.112045	11.5415	7.85651×10^{-21}

	Curvature
Max Intrinsic	0.648789
Max Parameter Effects	7.27395
95. % Confidence Region	0.703364

Figure 4.10. Output results of the nonlinear regression model (4.21) of *LPout*.

The third nonlinear model of *LPout* is of the following form, suggested partially by the model (4.15):

$$\begin{aligned}\widehat{LPout}(\theta,\lambda) &= \theta_0+\theta_1\psi_{YJ}(\lambda_1,F_1)+\theta_3\psi_{YJ}(\lambda_3,F_3)\\ &+\theta_4\left(\psi_{YJ}(\lambda_1,F_1)\right)^2\psi_{YJ}(\lambda_3,F_3)+\theta_5\psi_{YJ}(\lambda_1,F_1)\times \\ &\times\psi_{YJ}(\lambda_2,F_2)\psi_{YJ}(\lambda_3,F_3)+\theta_6\left(\psi_{YJ}(\lambda_3,F_3)\right)^3\end{aligned}\tag{4.22}$$

The coefficients of model (4.22) are seen in the output screen, shown in Figure 4.9. All necessary basic properties for validation of models (4.21) and (4.22) are satisfied.

Out[17]= 0.0839852

Out[18]=

	Estimate	Standard Error	t Statistic	P-Value
theta0	3.66158	0.0457444	80.0443	1.10096×10^{-100}
theta1	0.808217	0.0416437	19.4079	1.26833×10^{-37}
theta3	0.960674	0.081504	11.7868	2.41565×10^{-21}
theta5	−0.230984	0.0971224	−2.37828	0.0190868
theta6	−0.726854	0.0969714	−7.49555	1.65779×10^{-11}
theta7	−0.0217697	0.00738874	−2.94634	0.00391268
λ1	0.487396	0.0947546	5.14377	1.15664×10^{-6}
λ2	3.15287	0.721155	4.37197	0.0000276893
λ3	0.626162	0.209762	2.98511	0.00348151

Out[19]=

	Curvature
Max Intrinsic	0.402951
Max Parameter Effects	6.72374
95. % Confidence Region	0.713469

Figure 4.11. Output results of the nonlinear regression model (4.22) of *LPout*.

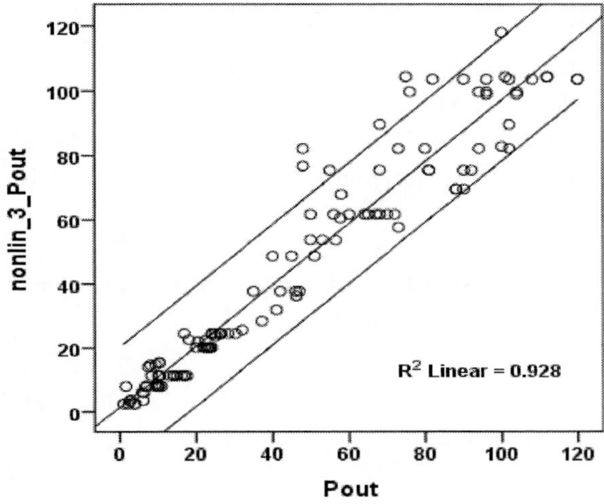

Figure 4.12. Comparison between the observed values of *Pout* and those calculated using the nonlinear model (4.22).

Figure 4.12 displays a comparative graph of experimental data on laser output power *Pout* against the transformed values of $\hat{P}out = Exp(\hat{L}Pout)$ calculated with the help of the nonlinear model (4.19). In particular, the highest experiment value of *Pout* = 120W is estimated by the model is 103.6W, which gives in this case 13 % relative error. F_1, F_2, F_3

Table 4.4. Statistics of the constructed nonlinear models of output power of CuBr lasers

Model number	*R*2	*R*2 adj.	Predictors in the model
(4.20)	0.900	0.899	F_1, F_2, F_3
(4.21)	0.915	0.914	$F_1, F_3, F_1^2 F_3, F_1 F_2 F_3$
(4.22)	0.928	0.927	$F_1, F_3, F_1^2 F_3, F_1 F_2 F_3, F_3^3$

4.4. Comparison of All Parametric Models

The summaries of all constructed polynomial and nonlinear models of laser efficiency are given in Table 4.5. This table indicates that the polynomial and nonlinear models display almost identical goodness of fit, presented by the coefficients of determination R^2 and adjusted R^2. Third degree models demonstrate the highest values. We have to add that complicating the model further with higher order members, yields worse indexes. At this stage, we can conclude that third degree polynomial and nonlinear models offer the best approximations of *Eff*.

Table 4.5. Comparison of the summaries of the parametric regression models of *Eff*

Parametric regression model	*R*2	*R*2 adj.	Predictors
Multiple Linear Model (3.4)	0.786	0.781	F_1, F_2, F_3
Polynomial first degree (4.4)	0.786	0.781	F_1, F_2, F_3, *PNE*
Polynomial second degree (4.5)	0.825	0.818	F_1, F_2, F_3, *PNE*, F_1^2
Polynomial third degree (4.6)	0.831	0.825	$F_1, F_3, F_1^2 F_3$, $F_1 F_2 F_3$, F_3^3, F_2^2, F_1^2, *PNE*
Nonlinear first degree (4.17)	0.790	0.788	F_1, F_2, F_3
Nonlinear second degree (4.18)	0.825	0.823	F_1, F_2, F_3, F_1^2
Nonlinear third degree (4.19)	0.839	0.837	F_1, F_2, F_3, $F_1^2 F_3$

Table 4.6. Comparison of the summaries of the parametric regression models of *LPout*

Parametric regression model	*R*2	*R*2 adj.	Predictors
Multiple Linear Model (3.6)	0.887	0.899	F_1, F_2, F_3
Polynomial first degree (4.4)	0.912	0.909	F_1, F_2, F_3
Polynomial second degree (4.5)	0.917	0.913	F_1, F_2, F_3, $F_1 F_2$, F_1^2
Polynomial third degree(4.6)	0.936	0.932	F_1, F_2, F_3, $F_1^2 F_3$
Nonlinear first degree (4.17)	0.900	0.899	F_1, F_2, F_3
Nonlinear second degree (4.18)	0.915	0.914	F_1, F_3, $F_1^2 F_3$, $F_1 F_2 F_3$
Nonlinear third degree (4.19)	0.928	0.927	F_1, F_3, $F_1^2 F_3$, $F_1 F_2 F_3$, F_3^3

Table 4.6 represents analogical summaries of the parametric regression models of *LPout.* It is seen that the third degree polynomial and nonlinear models have almost equal values of R^2 and adjusted R^2. We can conclude that third degree polynomial and nonlinear models give the best approximations of *LPout* and respectively of *Pout.*

In this section, polynomial and nonlinear regression models for the prediction of previous or future experiments are obtained. As a whole, we can qualify all polynomial and nonlinear models of the third degree as somewhat better than all parametric models of laser efficiency and output power of copper bromide laser.

REFERENCES

[1] S. G. Gocheva-Ilieva and I. P. Iliev, *Parametric and nonparametric empirical regression models of copper bromide laser generation, Math. Probl. Eng., Theory, Methods and Applications,* Hindawi Publ. Corp., New York, NY, Article ID 697687 (2010), 15 pages.

[2] S. G. Gocheva-Ilieva and I. P. Iliev, *Nonlinear regression model of copper bromide laser generation*, Proc. COMPSTAT'2010, eds. Y. Lechevallier, G. Saporta, 19th Int. Conf. Comp. Statistics, Paris - France, August 22-27, Springer_ebook, *Physica-Verlag*, 2010, 1063-1070.

[3] http://www.spss.com/software/statistics/stats-pro/, IBM SPSS Statistics.

[4] http://www.wolfram.com/mathematica/, Wolfram *Mathematica.*

[5] G. E. P. Box and D. R. Cox, *An analysis of transformations*, *Journal of the Royal Statistical Society*, Ser. B, 26 (1964) 211-252.

[6] K. Yeo and R. A. Johnson, A new family of power transformations to improve normality or symmetry, *Biometrika*, Oxford Press, 87(4) (2000) 954–959.

[7] http://reference.wolfram.com/mathematica/ref/NonlinearModelFit.html.

Chapter 5

MARS Models and Prediction of Laser Efficiency and Power of a Copper Bromide Vapor Laser

Abstract

In this Chapter, MARS models have been obtained based on all available data for examined CuBr lasers and not only on random samples from the conducted experiments. For this reason, the data are not random, since they have been selected by the researcher. On the other hand, in this way, the fullest possible information about the investigated dependences is utilized. Unlike classic parametric techniques, the models in this chapter are entirely data driven.

The data are taken from 274 observations of all 12 variables, described in Table 2.2. Observations where some measurements are missing have not been included.

All models are the best MARS models of the respective type.

Although the presented models are complex in form, they require few computations when calculating a predicted value. What is more, these are relatively easy to interpret for each specific experiment case. Further on, this will be illustrated by examples.

The goals in this chapter are:

- Construction of MARS models for estimation of laser efficiency
- Construction of MARS models for estimation of laser output power
- Application of the obtained models for predicting laser efficiency and output power for known and future experiments

All models have been obtained with the help of the MARS® software SPM (Salford Predictive Miner), version 6.6.0.083 [1].

5.1. Construction of MARS Models with the Help of Specialized Software

We will describe in brief the initial steps of working with the SPM-MARS software.

After starting the product, the data are loaded using the *File/Open/ DataFile* command. It is possible to load over 30 file formats, including ASCII (*.dat; *.txt; *.csv) and the common data files of the packages: Access, Excel, FoxPro, S-PLUS, Matlab, SAS, SPSS, Stata, Statistica, etc. With Excel files, the screen appears as shown in Figure 5.1.

After that, the command *Model/Construct Model* is selected from the main menu and the *Model Setup* dialog appears. The predictors and the response variable are chosen from *Variable Selection* (see Figure 5.2).

The *Options and Limits* box in the same window – *Model Setup* – is used to set:

- Maximum Basic Functions
- Maximum Interactions
- Models to Compute (it is preferable to select Best model only)
- Speed Factor (by default level 4)
- Penalty of added variables

and other options.

We select *Start* and the desired model is built. A separate window, similar to that shown in Figure 5.3 opens for each model. Detailed information about the elements and indices of the model can be called up by opening the boxes:

- *Summary* – as shown in Figure 5.3
- *ANOVA Decomp.* – statistical estimate of each BF (Basis Function)
- *Variable Importance* – relative variable importance in the model
- *Basis Functions* – BF and formula of the model
- *Gains*
- *Curves and Surfaces*
- *Final Model* – final model after backward stepwise elimination

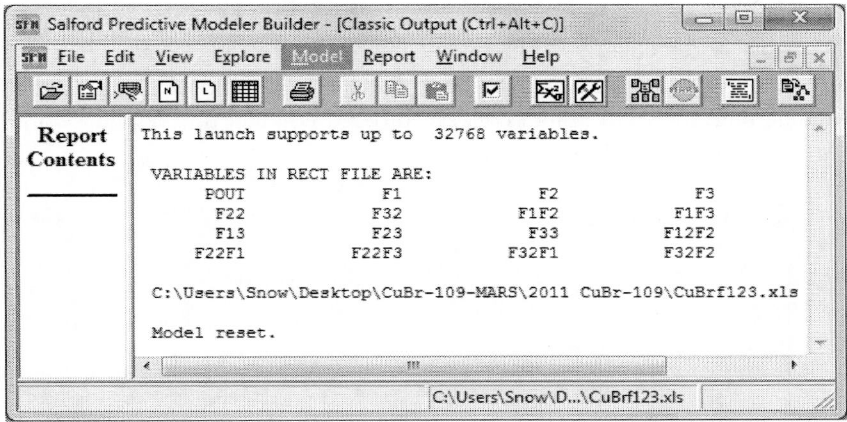

Figure 5.1. Main screen of the MARS software by the company Salford Predictive Modeler Builder [1].

The values of the dependent variable, predicted by the model are calculated using the *Score/Save results to a file/Select* command after entering a file name and selecting one of the input formats.

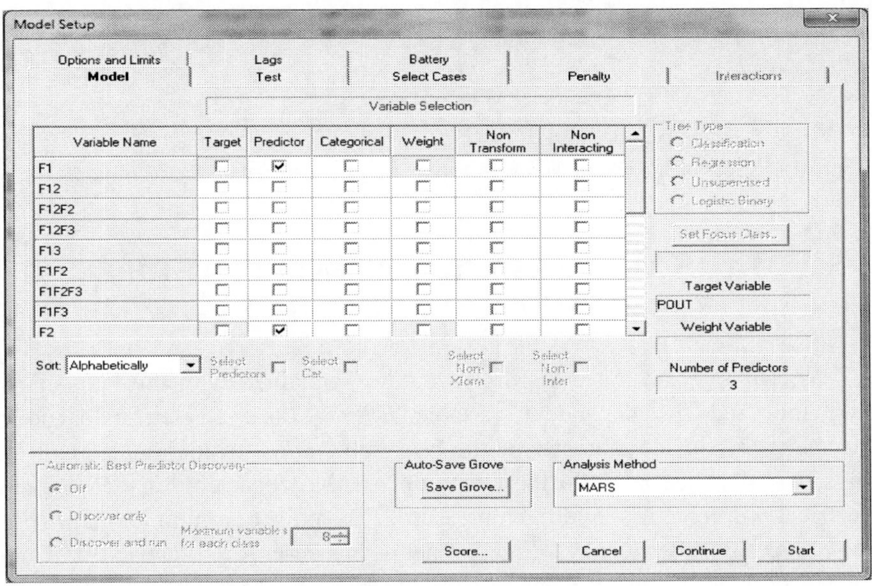

Figure 5.2. *Model Setup* window - *Variable Selection* dialog in MARS.

The software makes it possible to simultaneously generate and review in any order a practically unlimited number of models. Calculations are performed quickly in real time. A large number of additional options are also available but those will not be concerned here. You can find help on working with the product in [1,2].

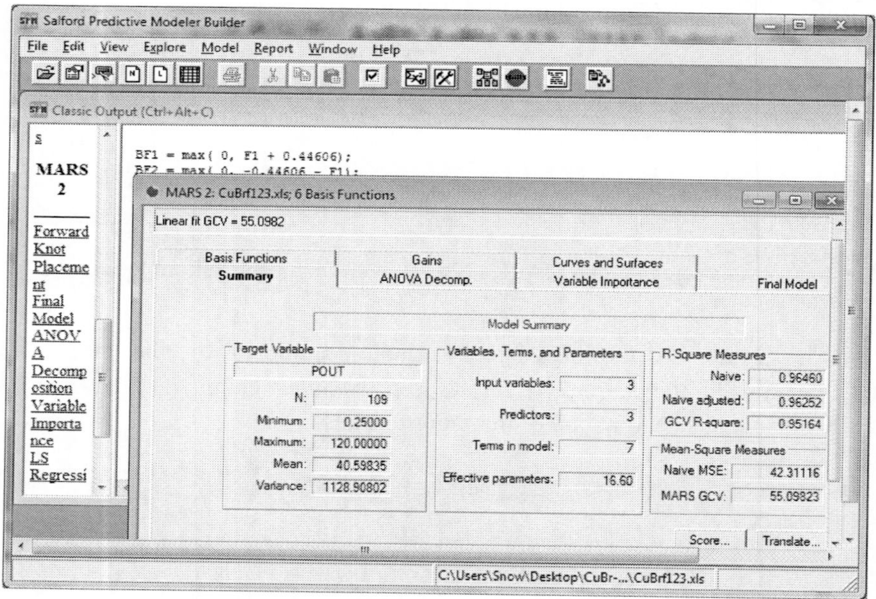

Figure 5.3. View of the results screen for the current constructed MARS model.

5.2. MARS Models with All Data for Modeling Laser Efficiency of CuBr Laser

We will present the best MARS models, obtained with the help of the predictors and the dependent variable *Eff*. We use all ten independent variables: *D* – inside diameter of the laser tube, *DR* – inside diameter of the rings, *L* – distance between the electrodes (active zone length), *PIN* – input electrical power, *PH2* – hydrogen pressure, *PL* – power per unit length, *PRF* – pulse repetition frequency, *PNE* – neon pressure, *C* – capacity of the condenser battery, and *TR* – temperature of the copper bromide reservoir.

The following types of MARS models have been constructed:

- of zero order – without interaction between predictors
- of first order – with possible interaction between predictors up to second degree members
- of second order – with possible interaction between predictors up to third degree members.

5.2.1. Zero Order Models of Laser Efficiency

These models include only piecewise linear non-interacting BF, derived from the given ten predictors.

Since the number of predictors is high (10), we will construct models with at least 2-3 times more BF. The summaries of the several models from this type are given in the common Table 5.1. We will consider in detail the model with relatively best indices – model 0-40, with up to 40 BF candidates for inclusion in the best MARS model.

We will adhere to the numbering of the functions as generated by the SPM software.

The model 0-40 includes nine predictors, in their order of importance: *C*, *PRF*, *DR*, *PH2*, *PIN*, *TR*, *PL*, *D*, and *PNE* (see also Table 5.2).

They are used in the following 22 BF:

$$
\begin{aligned}
&BF1 = \max(0,\ PIN-2)\\
&BF2 = \max(0,\ 2-PIN)\\
&BF3 = \max(0,\ DR-4.5)\\
&BF4 = \max(0,\ C-1.9)\\
&BF7 = \max(0,\ 0.3-PH2)\\
&BF8 = \max(0,\ PRF-16)\\
&BF10 = \max(0,\ PRF-21.5)\\
&BF12 = \max(0,\ PIN-2.3)\\
&BF14 = \max(0,\ PRF-14)\\
&BF16 = \max(0,\ D-50)\\
&BF19 = \max(0,\ 6.25-PL) \qquad\qquad (5.1)\\
&BF20 = \max(0,\ C-1.3)\\
&BF22 = \max(0,\ C-1.1)\\
&BF24 = \max(0,\ C-1)
\end{aligned}
$$

$$BF26 = \max(0,\ PRF - 18.5)$$
$$BF28 = \max(0,\ PRF - 23)$$
$$BF30 = \max(0,\ PL - 12)$$
$$BF33 = \max(0,\ 450 - TR)$$
$$BF34 = \max(0,\ D - 46)$$
$$BF36 = \max(0,\ PNE - 8)$$
$$BF37 = \max(0,\ PL - 11.25)$$
$$BF39 = \max(0,\ PL - 12.5)$$

The graphs of the functions in (5.1) are shown in Figure 5.4 (a)-(i). It is easy to directly read and compare how the changes in the nine laser input variables influence the behavior of the response *Eff* in pure ordinal units.

The estimated values of laser efficiency *Eff* could be calculated using the model equation:

$$\begin{aligned} \hat{Eff} = {} & -0.1465 + 2.6957\,BF1 + 0.5344\,BF2 + 0.0376\,BF3 \\ & +1.735BF4 - 1.1087\,BF7 - 0.2126\,BF8 + 0.3872\,BF10 \\ & -2.821\,BF12 + 0.188\,BF14 - 0.9047\,BF16 + 0.3647\,BF19 \\ & -4.442\,BF20 + 11.2647\,BF22 - 7.9737\,BF24 \\ & -0.1454\,BF26 - 0.2166\,BF28 - 1.5927\,BF30 \\ & -0.0235\,BF33 + 0.5965\,BF34 - 0.0054\,BF36 \\ & +0.6382\,BF37 + 0.9814\,BF39 \end{aligned} \tag{5.2}$$

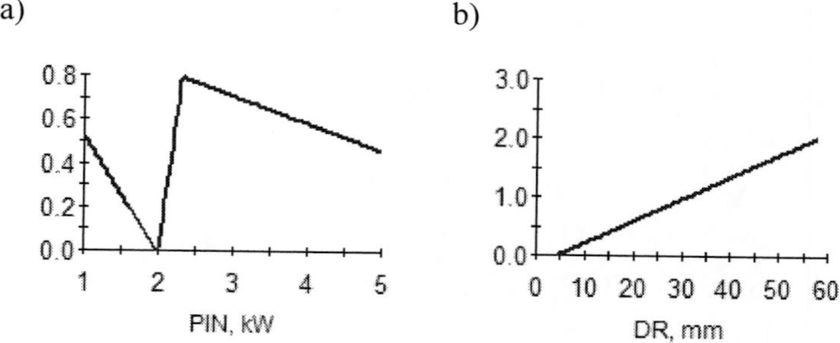

Figure 5.4. (Continued).

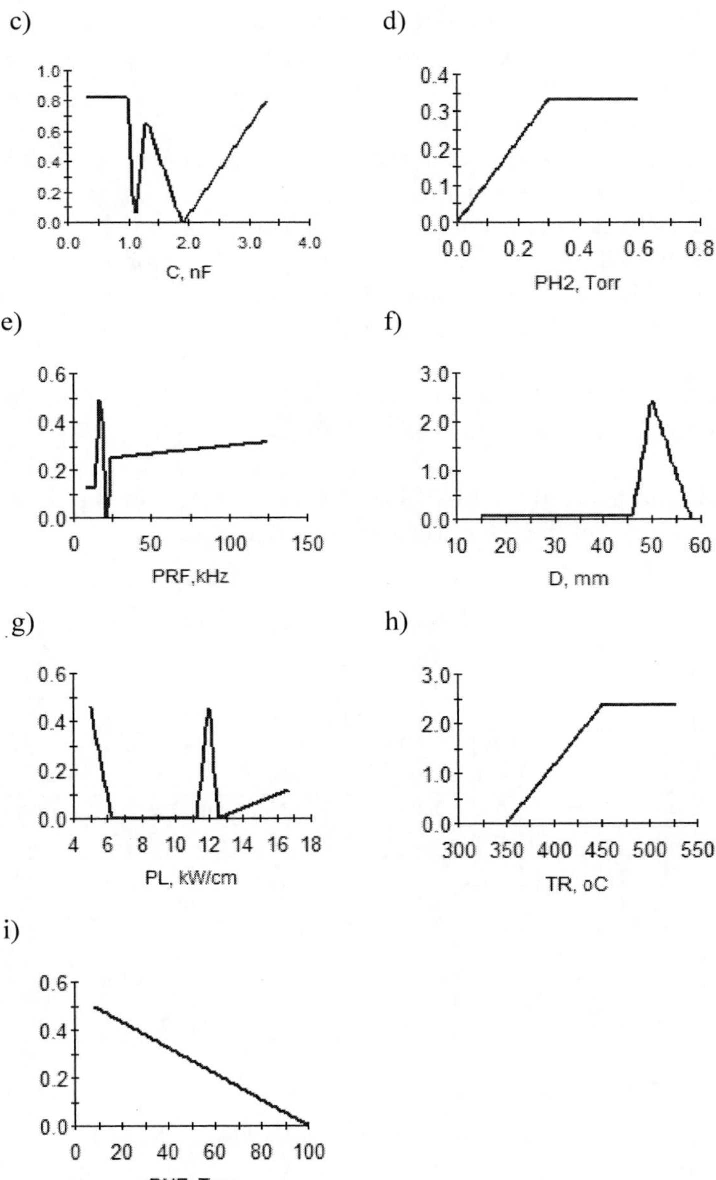

Figure 5.4. a)-i). Piecewise linear dependence between each of the nine predictors and *Eff* in MARS model (5.1)-(5.2) in pure ordinal units.

Model (5.1)-(5.2) is characterized by a coefficient of determination R-square = 0.961, Adj. R-square = 0.958, generalized cross validation measure GCV R-square=0.939, standard error of regression equals to 0.16350, and absolute deviation 0.12016. The model is statistically significant at the 0.1E-14 level. Its basic indices are given in Table 5.1.

Standard error of the estimate measures the dispersion of the dependent variable *Eff* estimate around its mean. Normally, if Std. Error is more than 10% of the mean, it is considered to be high. In our model (5.1)-(5.2) the mean of the predicted values is 1.551. So, we obtain a relative error of about 10%.

The model is tested by generating the best MARS model, selected so as to allow no overfitting of the model and with the best GCV, as well as by using the ordinary least squares method.

The relative variable importance is given in Table 5.2 (in MARS the most important variable always has a value of 100).

Table 5.1. Summary of the MARS models constructed for the estimation of the laser efficiency of a CuBr laser

Model type	Max BF	*R2*	*R2* adj.	MARS GCV *R2*	Stand. Error of the Estimate (SEE)	Abs. Deviation	%	Predictors / BF in the model
0th order	25	0.947	0.944	0.930	0.1905	0.1427		7/15
	30	0.954	0.950	0.934	0.1786	0.1318		8/18
	40	**0.961**	**0.958**	**0.939**	**0.1635**	**0.1202**	**11**	**9/22**
	50	0.961	0.958	0.938	0.1635	0.1202		9/22
1st order	30	0.972	0.970	0.956	0.1389	0.0990		6/20
	40	0.979	0.977	0.965	0.1204	0.0895		8/22
	50	**0.983**	**0.981**	**0.965**	**0.1086**	**0.0801**	7	**9/28**
2nd order	30	0.967	0.965	0.952	0.1508	0.1109		6/16
	40	0.983	0.981	0.966	0.1085	0.0809		8/29
	50	**0.984**	**0.983**	**0.970**	**0.1037**	**0.0773**	7	**7/28**

With the help of MARS model (5.1)-(5.2) it is easy to calculate the estimate values $\widehat{Eff}$ when predictor values are known. The same is valid for predicting a future response. For example, a maximum laser efficiency *Eff* = 3.07% has been measured at *D*=58 mm, *DR*=58 mm, *L*=200 cm, *PIN*=5 kW, *PL*=7.5 kW/cm (with 50% losses), *PH2*=0.6 Torr, *PRF*=17.5 kHz, *PNE*=20

Torr, *C*=1 nF, and *TR*=490 °C [3, 4]. After substituting the latter in (5.1)-(5.2) we find $\widehat{Eff} \approx 2.787\%$.

Table 5.2. Relative variable importance in the best MARS models of laser efficiency

Variable	Importance in the model		
	Model (5.1)-(5.2)	Model (5.3)-(5.4)	Model (5.5)-(5.6)
D	18	14	-
DR	49	76	30
L	-	40	-
PIN	33	70	100
PL	20	46	16
PH2	40	37	20
PRF	67	71	29
PNE	3	-	-
C	100	100	44
TR	22	20	12

5.2.2. First Order Models of Laser Efficiency

The second type of MARS models of *Eff* using all data demonstrates possible piecewise first order interactions. Therefore, these models may include BF with the products of the predictors.

The results from some of the initial constructed first order models are given in Table 5.1. As a whole, they are relatively close. We will present a model with the minimum relative percent of Std. Error. Model 1-50, which employs the following 30 BF, is shown of the following page.

$$
\begin{aligned}
BF1 &= \max(0,\ PIN-2) \\
BF2 &= \max(0,\ 2-PIN) \\
BF3 &= \max(0,\ C-0.33)\,BF1 \\
BF5 &= \max(0,\ 0.36-PH2)\,BF2 \\
BF6 &= \max(0,\ PRF-17)\,BF1 \\
BF7 &= \max(0,\ 17-PRF)\,BF1 \\
BF8 &= \max(0,\ DR-4.5) \\
BF9 &= \max(0,\ C-1.9)\,BF8 \\
BF10 &= \max(0,\ 1.9-C)\,BF8 \\
BF13 &= \max(0,\ PRF-16)\,BF8 \\
BF14 &= \max(0,\ 16-PRF)\,BF8 \\
BF15 &= \max(0,\ C-1.1) \\
BF16 &= \max(0,\ 1.1-C) \\
BF17 &= \max(0,\ PIN-4)\,BF15 \\
BF18 &= \max(0,\ 4-PIN)\,BF15 \\
BF21 &= \max(0,\ PRF-18.5)\,BF1 \\
BF23 &= \max(0,\ PIN-1.2)\,BF16 \\
BF24 &= \max(0,\ 1.2-PIN)\,BF16 \\
BF25 &= \max(0,\ PIN-1.4)\,BF16 \\
BF30 &= \max(0,\ 480-TR)\,BF2 \\
BF32 &= \max(0,\ 7.5-PL)\,BF2 \\
BF34 &= \max(0,\ 120-L)\,BF8 \\
BF36 &= \max(0,\ 1.3-C) \\
BF37 &= \max(0,\ PIN-2.3)\,BF36 \\
BF38 &= \max(0,\ 2.3-PIN)\,BF36 \\
BF41 &= \max(0,\ 8.75-PL)\,BF1 \\
BF44 &= \max(0,\ PRF-17.5)\,BF2 \\
BF48 &= \max(0,\ PRF-17.5) \\
BF49 &= \max(0,\ 17.5-PRF) \\
BF50 &= \max(0,\ D-15)\,BF48
\end{aligned}
\tag{5.3}
$$

The corresponding regression equation used to calculate the predicted values of *Eff* is

$$\begin{aligned}\hat{Eff} =\ & 0.267 - 11.574\ BF1 + 14.602\ BF3 - 1.317\ BF5 \\ & + 0.0766\ BF6 + 0.0466\ BF7 + 0.2496\ BF8 + 0.306\ BF9 \\ & - 0.241\ BF10 + 0.0017\ BF13 - 0.0078\ BF14 \\ & - 43.136\ BF15 + 12.013\ BF16 - 14.491\ BF17 \\ & + 13.763\ BF18 - 0.125\ BF21 - 45.199\ BF23 \\ & - 20.19\ BF24 + 54.155\ BF25 - 0.01037\ BF30 \\ & + 1.46\ BF32 + 0.00057\ BF34 + 3.37\ BF37 \\ & - 4.596\ BF38 + 0.142\ BF41 + 0.29\ BF44 \\ & - 0.2877\ BF48 + 0.0799\ BF49 + 0.0009\ BF50\end{aligned} \tag{5.4}$$

Using (5.3)-(5.4) for the maximum laser efficiency 3.07%, we obtain $\hat{Eff} \approx 2.995\%$ (compare with 5.1.1).

Figure 5.5 shows the graphs of the contribution of the first order interactions in equation (5.4).

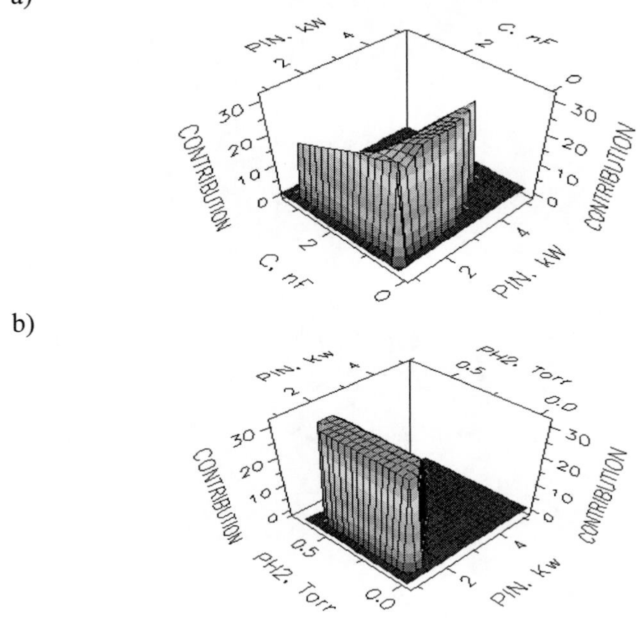

Figure 5.5. (Continued).

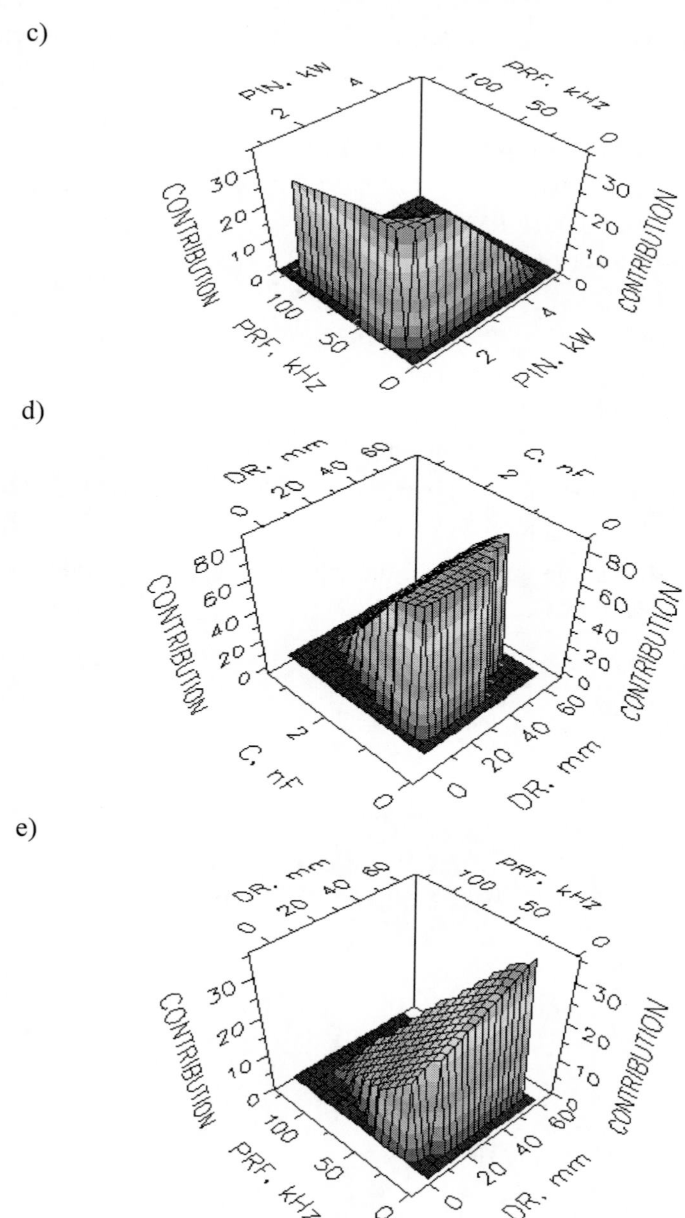

Figure 5.5. (Continued).

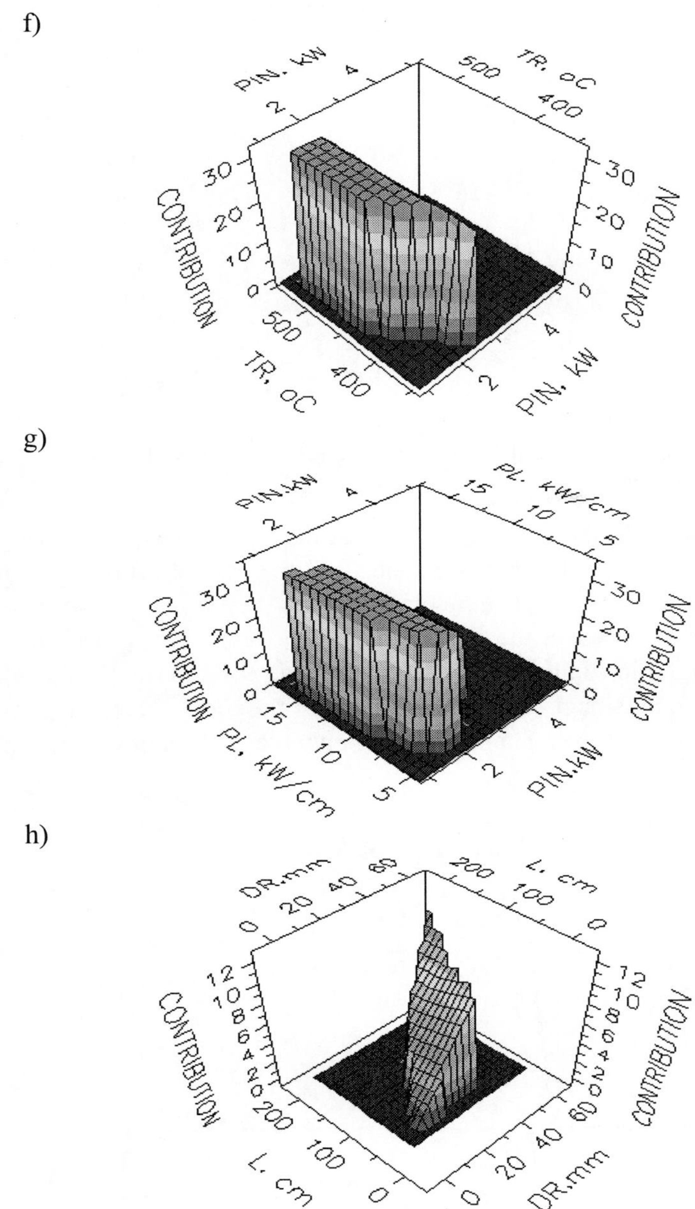

Figure 5.5. (Continued).

i)

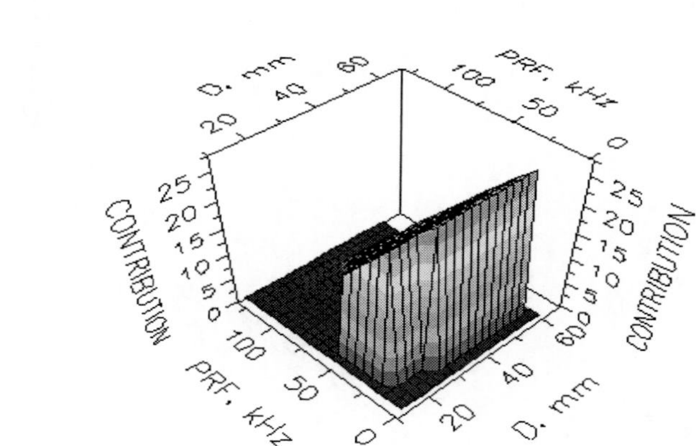

Figure 5.5. a)-i). Combined contributions of predictors to the values of *Eff* from model (5.3)-(5.4) in ordinal units.

The analysis of these graphs shows that the biggest contribution (up to a maximum of 80 ordinal units) is provided by the mutual influence of the variables *DR* and *C*. In (5.3) this corresponds to basis functions BF9 and BF10. We are interested in the case of large-bore lasers, for which *DR*>40 mm. For example, in Figure 5.6 it easy to observe the behavior of *C* when *DR*=58 mm. The maximum is reached when C is close to 1. Therefore, when constructing new lasers with similar *DR*, it is suggested to begin with *C* in the interval [1, 2].

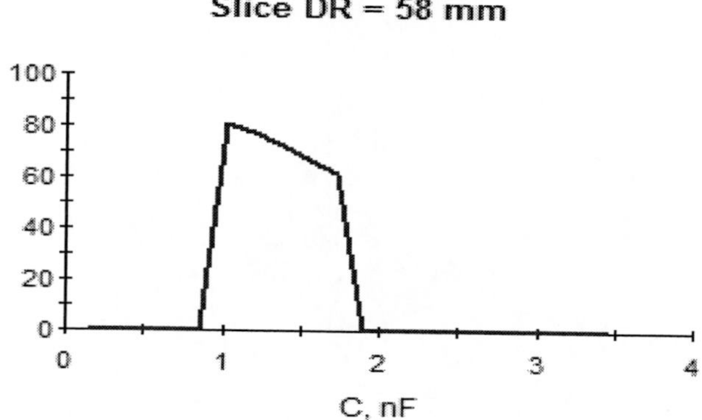

Figure 5.6. Slice from Figure 5.5d with the element of the graphics for *DR*=58 mm.

5.2.3. Second Order Models of Laser Efficiency

The next type of MARS models is this with possible piecewise second order interactions. Their BF may contain products of three linear members.

Some of the obtained models of second order are shown in Table 5.1. Relatively the best one is model 2-50. Its relative standard error is less than 7%.

For a preset maximum of 50 basis functions, the best MARS model of this type includes the following 31 basis functions:

$$
\begin{aligned}
&BF1 = \max(0,\ PIN-2)\\
&BF2 = \max(0,\ 2-PIN)\\
&BF3 = \max(0,\ C-0.33)\,BF1\\
&BF4 = \max(0,\ PH2-0.36)\,BF2\\
&BF5 = \max(0,\ 0.36-PH2)\,BF2\\
&BF6 = \max(0,\ PRF-17)\,BF1\\
&BF7 = \max(0,\ 17-PRF)\,BF1\\
&BF8 = \max(0,\ DR-4.5)\\
&BF10 = \max(0,\ 1.9-C)\,BF8\\
&BF11 = \max(0,\ PIN-3)\,BF10\\
&BF12 = \max(0,\ 3-PIN)\,BF10\\
&BF13 = \max(0,\ PRF-15.5)\,BF10\\
&BF14 = \max(0,\ 15.5-PRF)\,BF10\\
&BF15 = \max(0,\ C-1.3)\,BF6\\
&BF17 = \max(0,\ C-1)\,BF5\\
&BF18 = \max(0,\ 1-C)\,BF5\\
&BF20 = \max(0,\ 480-TR)\,BF5\\
&BF21 = \max(0,\ PIN-1)\,BF8\\
&BF23 = \max(0,\ 18.5-PRF)\,BF21\\
&BF30 = \max(0,\ C-1.1)\,BF2\\
&BF31 = \max(0,\ 1.1-C)\,BF2\\
&BF32 = \max(0,\ C-1.3)\\
&BF34 = \max(0,\ PL-12)\,BF2\\
&BF35 = \max(0,\ 12-PL)\,BF2
\end{aligned}
\qquad (5.5)
$$

$$BF36 = \max(0,\ PL - 7.5)\, BF2$$
$$BF38 = \max(0,\ PRF - 17)\, BF21$$
$$BF40 = \max(0,\ PRF - 17.5)\, BF8$$
$$BF43 = \max(0,\ 1.1 - C)\, BF40$$
$$BF45 = \max(0,\ 485 - TR)\, BF36$$
$$BF46 = \max(0,\ PRF - 18.5)\, BF3$$
$$BF48 = \max(0,\ PL - 10)\, BF32$$

The corresponding regression equation for predicting values of *Eff* is

$$\begin{aligned}\widehat{Eff} = {} & 1.086 - 9.347\, BF2 - 0.731\, BF3 - 1.843\, BF4 - 3.0463\, BF5 \\ & + 0.185\, BF6 + 0.223\, BF7 - 0.0292\, BF11 + 0.008\, BF12 \\ & - 0.0025\, BF13 - 0.00397\, BF14 - 0.129\, BF15 + 18.183\, BF17 \\ & - 86.855\, BF18 - 0.755\, BF20 + 0.0313\, BF21 - 0.0039\, BF23 \\ & + 1.224\, BF30 + 4.941\, BF31 - 1.388\, BF32 - 2.604\, BF34 \\ & + 1.519\, BF35 + 1.9405\, BF36 - 0.00652\, BF38 + 0.0032\, BF40 \\ & + 0.017\, BF43 + 0.0072\, BF45 + 0.1503\, BF46 - 0.229\, BF48 \end{aligned} \quad (5.6)$$

It needs to be noted that 14 BF are of third degree and 13 are of the second degree. Only 4 BF are piecewise linear. These means that model (5.5)-(5.6) contains essentially nonlinear terms.

There are no graphic results for this model, indicating the contribution of the interactions at hand. The importance of the predictors in the regression equation (5.6) is given in Table 5.2.

For example, for the same case of a maximum laser efficiency *Eff* = 3.07%, given in 5.1.1, model (5.5)-(5.6) predicts a value $\widehat{Eff} \approx 3.036\%$.

Figure 5.7 displays the predicted values of *Eff* calculated using the second order model (5.5)-(5.6) and the corresponding measured values of laser efficiency (with 5% confidence interval).

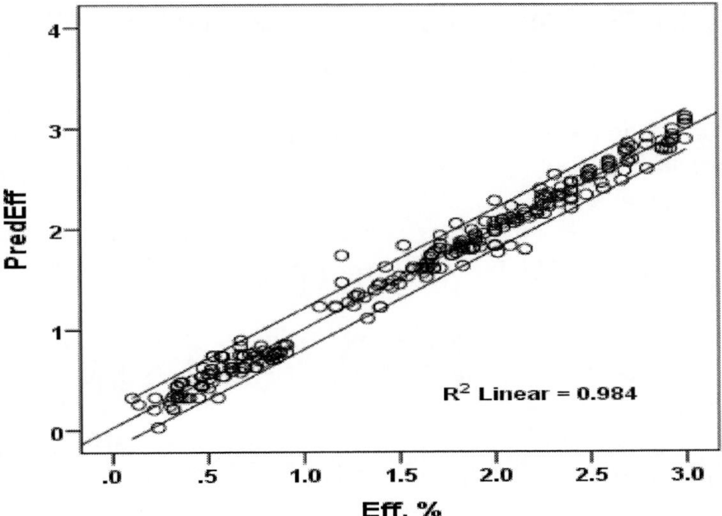

Figure 5.7. Comparative graph of the values predicted by the best MARS model (5.5)-(5.6) versus the measured values of laser efficiency *Eff*.

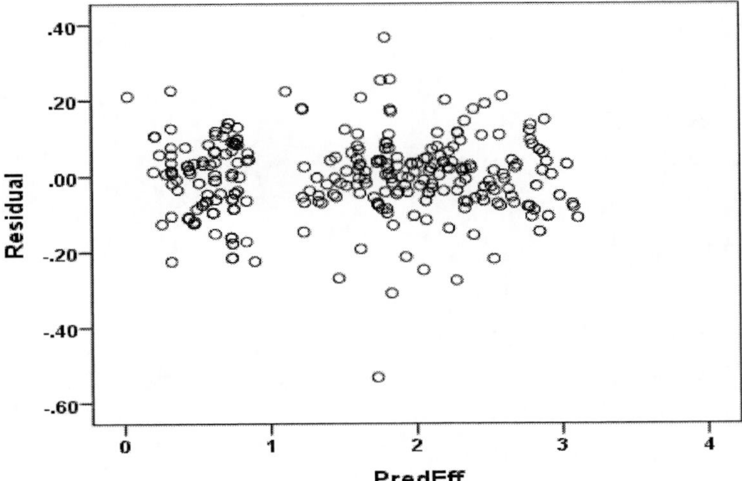

Figure 5.8. Investigation of the homoscedasticity of model (5.5)-(5.6).

The quality of the approximation with model (5.5)-(5.6) is presented in Figure 5.8 by the plot of the residuals against the predicted values of the dependent variable *Eff*. It is obvious that there is no dependence. Therefore, the heteroscedasticity effect is absent.

5.3. MARS Models of Output Power of a Copper Bromide Vapor Laser

In this section, we will describe some of the models, built for predicting the output power *Pout*.

5.3.1. Zero Order Models of Output Power

Of the models from this type, given in Table 5.3, we will only describe the zero order model with 30 BF. Within it, the following 16 BF have been generated:

$$
\begin{aligned}
&BF1 = \max(0,\ PIN - 2.5)\\
&BF3 = \max(0,\ C - 1.9)\\
&BF5 = \max(0,\ DR - 40)\\
&BF6 = \max(0,\ 40 - DR)\\
&BF7 = \max(0,\ C - 1.3)\\
&BF9 = \max(0,\ PIN - 3)\\
&BF11 = \max(0,\ PRF - 16)\\
&BF13 = \max(0,\ PRF - 21.5)\\
&BF15 = \max(0,\ PRF - 14)\\
&BF17 = \max(0,\ PIN - 2)\\
&BF19 = \max(0,\ PRF - 18.5)\\
&BF21 = \max(0,\ PRF - 23)\\
&BF25 = \max(0,\ C - 1.1)\\
&BF27 = \max(0,\ C - 1)\\
&BF29 = \max(0,\ PL - 10.4167)\\
&BF30 = \max(0,\ 10.4167 - PL)
\end{aligned}
\qquad (5.7)
$$

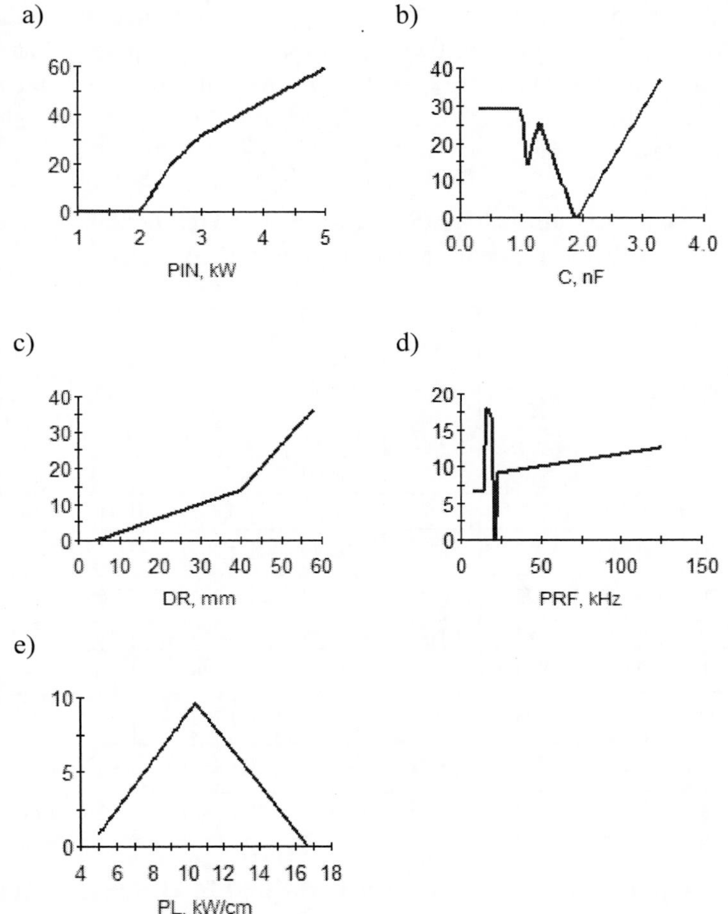

Figure 5.9. a)-e). Contribution of the predictors to *Pout* in model (5.7)-(5-8).

The corresponding regression equation for predicting the values of *Pout* is:

$$\begin{aligned}Pout = 26.1898 - 15.077\ BF1 + 70.773\ BF3 + 1.237\ BF5 \\ - 0.393 BF6 - 106.966\ BF7 - 10.1247\ BF9 - 5.47\ BF11 \\ + 15.187\ BF13 + 5.5869\ BF15 + 39.027\ BF17 \\ - 7.179\ BF19 - 8.09\ BF21 + 220.41\ BF25 \\ - 157.43\ BF27 - 1.541\ BF29 - 1.624\ BF30\end{aligned} \quad (5.8)$$

Model (5.7)-(5.8) includes the following 5 piecewise linear predictors: *C*, *PRF*, *PIN*, *DR*, *PL*. The plots of their influence on *Pout* in pure ordinal units are given in Figure 5.9. The first Figure 5.9a shows that when the variable *PIN* increases, *Pout* also increases. The dependence on *DR* – Figure5.9c is similar, although to a lesser degree.

Table 5.3. Summary of the MARS models constructed for the estimation of the laser output power Pout of a CuBr laser

Model type	Max BF	*R2*	*R2* adj.	MARS GCV *R2*	Stand. Error of the Estimate (SEE)	Abs. Deviation	%	Predictors / BF in the model
0th order	**30**	**0.980**	**0.979**	**0.973**	**5.1317**	**3.8810**	**11**	**5/16**
	40	0.980	0.978	0.973	5.2004	4.0027		6/15
	50	0.980	0.979	0.972	5.2004	4.0027		6/16
1st order	**30**	**0.994**	**0.994**	**0.991**	**2.78069**	**2.05019**	**6**	**6/21**
	40	0.996	0.995	0.992	2.3969	1.8285		7/28
	50	0.996	0.995	0.993	2.2948	1.7260		7/28
2nd order	**30**	**0.996**	**0.995**	**0.992**	**2.3993**	**1.8144**	**5**	**6/24**
	40	0.997	0.996	0.994	2.1198	1.6087		7/26

Table 5.4. Relative variable importance in the best MARS models of laser output power of a CuBr laser

Variable	Importance in the model		
	Model (5.7)-(5.8)	Model (5.9)-(5.10)	Model (5.11)-(5.12)
D	-	-	-
DR	44	97	93
L	-	-	-
PIN	67	100	100
PL	11	20	5
PH2	-	10	17
PRF	75	65	68
PNE	-	-	-
C	100.	77	87
TR	-	-	-

5.3.2. First Order Models of Output Power

Of the first order models the model with up to 30 BF. The model involves the following 22 BF:

$$
\begin{aligned}
&BF1 = \max(0,\ PIN - 2.5)\\
&BF2 = \max(0,\ 2.5 - PIN)\\
&BF5 = \max(0,\ 17.5 - PRF)\,BF1\\
&BF6 = \max(0,\ DR - 30)\\
&BF7 = \max(0,\ 30 - DR)\\
&BF8 = \max(0,\ C - 1.3)\,BF6\\
&BF9 = \max(0,\ 1.3 - C)\,BF6\\
&BF10 = \max(0,\ PRF - 16.5)\,BF6\\
&BF11 = \max(0,\ 16.5 - PRF)\,BF6\\
&BF12 = \max(0,\ PIN - 3)\,BF6\\
&BF13 = \max(0,\ 3 - PIN)\,BF6\\
&BF14 = \max(0,\ PH2 - 4.563E - 09)\,BF2\\
&BF15 = \max(0,\ PRF - 18.5)\,BF1\\
&BF17 = \max(0,\ PL - 8.21429)\,BF2\\
&BF18 = \max(0,\ 8.21429 - PL)\,BF2\\
&BF20 = \max(0,\ 13 - PL)\,BF7\\
&BF21 = \max(0,\ PL - 8.75)\,BF1\\
&BF22 = \max(0,\ 8.75 - PL)\,BF1\\
&BF23 = \max(0,\ C - 1.1)\,BF7\\
&BF26 = \max(0,\ 14.3 - PRF)\,BF7\\
&BF27 = \max(0,\ PRF - 21.5)\,BF6\\
&BF29 = \max(0,\ PRF - 20)\,BF6
\end{aligned}
\qquad (5.9)
$$

The corresponding regression equation for predicting the values of *Pout* includes 21 BF and has the form:

$$\begin{aligned}\hat{P}out = {} & 51.038 - 46.628\, BF2 - 0.9446\, BF5 + 1.0276\, BF6 \\ & - 1.117\, BF7 - 1.6807\, BF8 + 0.697 BF9 - 0.08578\, BF10 \\ & - 0.2297\, BF11 + 1.17\, BF12 - 0.875\, BF13 \\ & + 9.104\, BF14 - 3.403\, BF15 + 3.686\, BF17 + 9.005\, BF18 \\ & + 0.8086\, BF20 - 2.46\, BF21 + 11.002 BF22 \\ & + 0.321\, BF23 - 0.125\, BF26 + 0.284\, BF27 - 0.173\, BF29\end{aligned} \qquad (5.10)$$

In model (5.9)-(5.10) the initial six predictors are: *PIN, DR, PH2, PRF, PL* and *C*. Piecewise second degree interacting terms included in the model in decreasing order of their importance are: {*PIN.DR*}, {*PIN.PRF*}, {*PIN.PH2*}, {*PIN.PL*}, {*DR.C*}, {*PL.DR*} and {*DR.PRF*}. The regions with a clearly defined contribution to these influences can be observed in Figure 5.10.

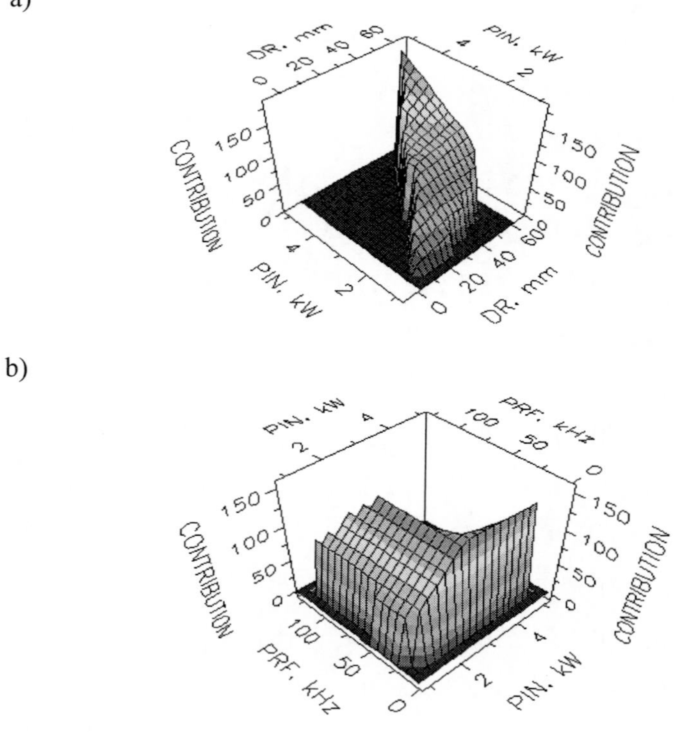

Figure 5.10. (Continued).

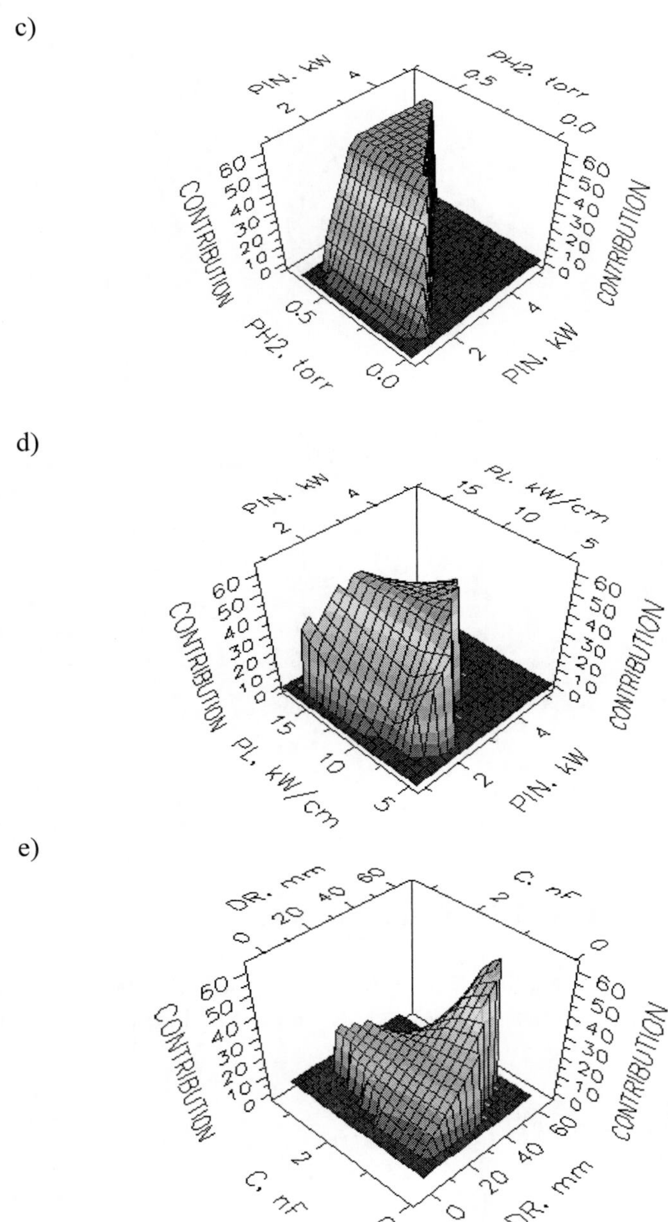

Figure 5.10. (Continued).

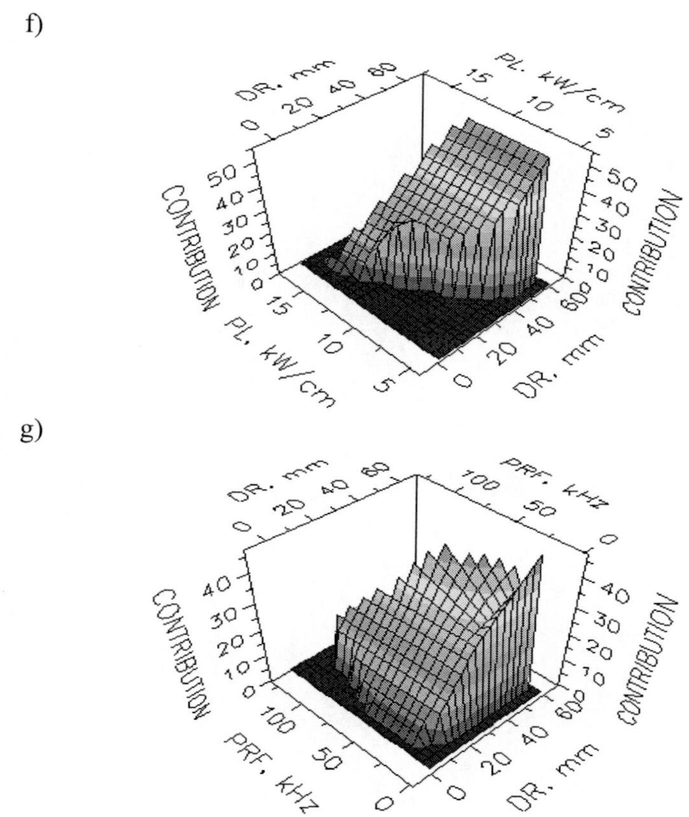

Figure 5.10. a)-g). Contributions of the predictor variables to the first order model of *Pout*, 30 BF.

5.3.3. Second Order Models of Output Power

Of the second order models in Table 5.3 we consider the model generated with 30 initial basis functions. The best MARS model includes the following 26 BF:

$$\begin{aligned} &BF1 = \max(0,\ PIN - 2.5) \\ &BF2 = \max(0,\ 2.5 - PIN) \\ &BF3 = \max(0,\ C - 0.333)\,BF1 \end{aligned} \qquad (5.11)$$

$$BF4 = \max(0,\ PRF - 17.5)\,BF1$$
$$BF5 = \max(0,\ 17.5 - PRF)\,BF1$$
$$BF6 = \max(0,\ DR - 30)$$
$$BF7 = \max(0,\ 30 - DR)$$
$$BF8 = \max(0,\ C - 1.3)\,BF6$$
$$BF9 = \max(0,\ 1.3 - C)\,BF6$$
$$BF10 = \max(0,\ PRF - 16.5)\,BF6$$
$$BF11 = \max(0,\ 16.5 - PRF)\,BF6$$
$$BF12 = \max(0,\ PIN - 1.5)\,BF9$$
$$BF13 = \max(0,\ 1.5 - PIN)\,BF9$$
$$BF14 = \max(0,\ PRF - 18.5)\,BF1$$
$$BF16 = \max(0,\ PH2 - 0.5)\,BF2$$
$$BF17 = \max(0,\ 0.5 - PRF)\,BF2$$
$$BF18 = \max(0,\ PL - 8.75)\,BF3$$
$$BF19 = \max(0,\ 8.75 - PL)\,BF3$$
$$BF21 = \max(0,\ 2 - PIN)\,BF10$$
$$BF22 = \max(0,\ PRF - 18.5)\,BF3$$
$$BF23 = \max(0,\ 18.5 - PRF)\,BF3$$
$$BF24 = \max(0,\ PRF - 20)\,BF6$$
$$BF26 = \max(0,\ PIN - 4)\,BF8$$
$$BF28 = \max(0,\ C - 1.1)$$
$$BF29 = \max(0,\ 1.1 - C)$$
$$BF30 = \max(0,\ C - 0.33)\,BF17$$

The regression equation for predicting the values of *Pout* has the form:

$$\begin{aligned} \hat{P}out = {} & 46.98 - 23.6048\,BF2 + 22.594\,BF3 + 2.953\,BF4 \\ & - 2.661\,BF5 + 0.92\,BF6 - 0.534\,BF7 - 3\,BF8 \\ & - 1.582\,BF9 - 0.108\,BF10 - 0.241\,BF11 + 1.677\,BF12 \\ & + 40.381\,BF13 - 10.489\,BF14 - 128.349\,BF16 \\ & - 0.9479\,BF18 + 8.219\,BF19 + 0.264\,BF21 \\ & + 3.358\,BF22 + 1.487\,BF23 - 0.122\,BF24 \\ & - BF26 + 6.737\,BF28 + 9.649\,BF29 - 11.703\,BF30 \end{aligned} \tag{5.12}$$

In model (5.11)-(5.12) the predictors are: *PIN*, *DR*, *PRF*, *C*, *PH2*, and *PL*. Piecewise second degree interacting terms included in the model are: {*PIN.PRF*}, {*PIN.PH2*}, {*PIN.PL*}, {*PIN.C*}, {*DR.PRF*} and {*DR.C*}. When accounting for the interactions of the second order (third degree products between the initial predictors), from (5.11) we have the following terms: {*PIN.PRF.C*}, {*PIN.PRF.DR*}, {*PIN.DR.C*}, and {*PIN.C.PL*}.

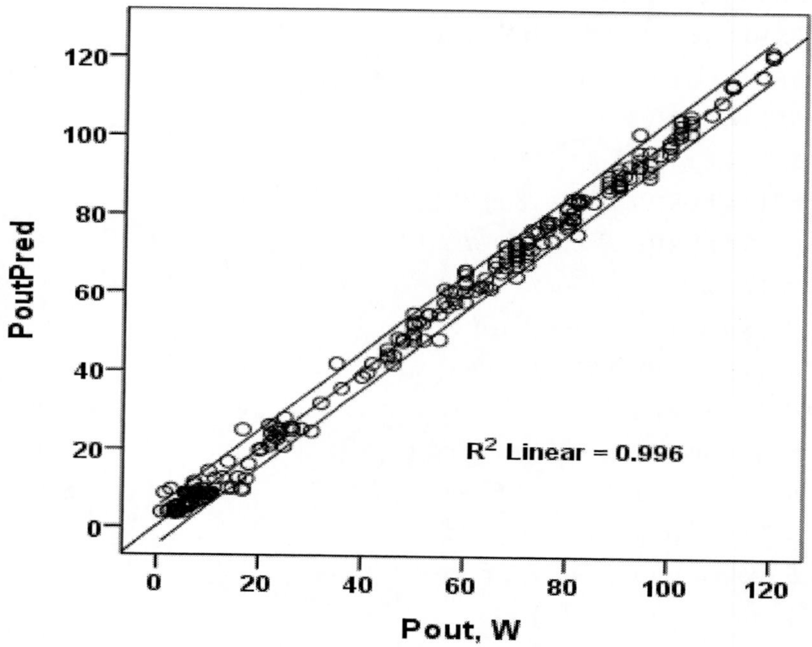

Figure 5.11. Comparative graph of the values predicted by the best MARS model (5.11)-(5.12) versus the measured values of laser output power *Pout*.

Figure 5.11 displays the predicted values of *Pout* calculated using the second order model (5.11)-(5.12) and the corresponding measured values (with 5% confidence interval).

The lack of heteroscedasticity effect in model (5.11)-(5.12) is seen from Figure 5.12.

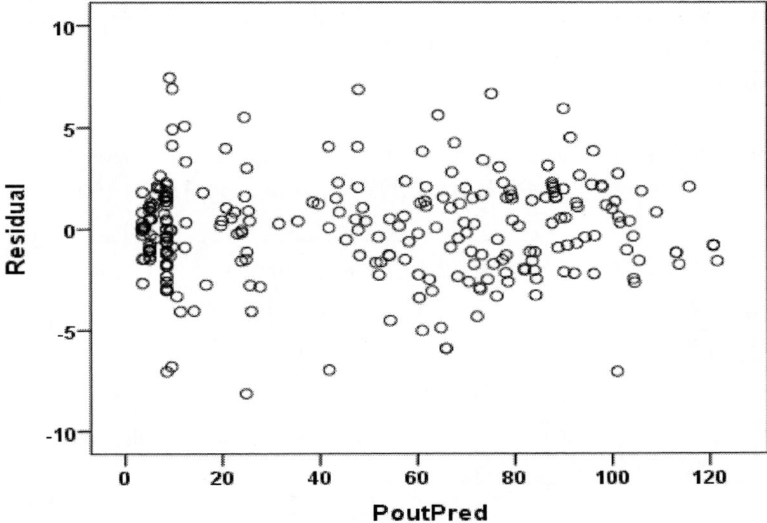

Figure 5.12. No heteroscedasticity in model (5.11)-(5.12) is observed.

5.4. Prediction of Laser Efficiency and Power Using MARS Models

This section deals with the application of the constructed nonparametric models of laser efficiency and output power of CuBr lasers. As shown in Tables 5.1 and 5.3, all MARS models are characterized by high statistical indices. Of these, here, we will use only the models with second order interactions between predictors. These models demonstrate relative standard errors within 5 to7%, which is comparable with the experiment error.

Our ultimate goal is to predict future experiments for lasers with increased output power and if possible – with increased efficiency. Since the selection of the independent variables for a future experiment is a difficult task, we will preset some established optimal values of laser characteristics and modify the basic ones, which are included in the considered model.

5.4.1. Prediction of Laser Efficiency

The results of the prediction are given in Table 5.5. The first row is the 'vmax' variant, which has the highest efficiency within the investigated dataset. The next 3 rows (v-2, v-1 and v0) are also actual data, presented for the sake of comparison. The measured values of *Eff* for these four cases are respectively: 3.07%, 2.49%, 2.49% and 2.4%.

Table 5.5. Variants of the presumed future experiments and predicted values of laser efficiency *EffPred* using model (5.5)-(5.6)*

Variant	*D*	*DR*	*L*	*PIN*	*PL*	*PH2*	*PRF*	*PNE*	*C*	*TR*	*Eff Pred*
vmax	58	58	200	3	7.5	0.6	17.5	20	1	490	3.036
v-2	58	58	200	4.5	11.25	0.6	16.5	20	1.3	490	2.517
v-1	58	58	200	4.5	11.25	0.6	17.5	20	1.3	490	2.503
v0	58	58	200	5	12.5	0.6	17.5	20	1.3	490	2.372
v1	58	58	200	5.05	12.63	0.6	17.5	20	1.3	490	2.359
v2	60	60	210	5.1	12.14	0.6	17.5	20	1.3	490	2.456
v3	65	63	215	5.15	11.98	0.6	16.5	20	1.3	490	2.631
v4	68	65	220	5.2	11.82	0.6	17.5	20	1.3	500	2.740
v5	68	65	220	5.3	12.05	0.65	18	20	1.1	500	2.325
v6	70	70	240	5.3	11.04	0.65	18	20	1.1	500	2.531
v7	70	70	220	5.3	12.05	0.65	18	20	1.1	500	2.531
v8	70	70	230	5.3	11.52	0.65	18	20	1.1	500	2.531
v9	70	70	230	5.3	11.52	0.65	18	20	1.2	500	2.770
v10	70	70	240	5.3	11.04	0.65	18	20	1.2	500	2.770
v11	70	68	240	5.3	11.04	0.6	17.5	20	1.3	500	2.900
v12	70	70	240	5.5	11.46	0.65	18	20	1.3	500	2.973
v13	70	70	240	5.5	11.46	0.6	17.5	20	1.3	500	2.985
v14	70	68	225	5.25	11.67	0.6	17.5	20	1.3	500	2.910
v15	75	73	230	5.3	11.72	0.6	17.5	20	1.2	500	2.951
v16	75	73	230	5.3	11.72	0.6	17.5	20	1.3	500	3.203

* The first four rows contain data from actual experiments.

The results presented in Table 5.5 show that when *DR*, *L* and *PIN* are increased (and the condition that *PL* decreases is satisfied), the expected improvement of laser efficiency can be observed. Naturally, if the future experiment deviates significantly from the actual data, the error will increase.

Each of the developed models can be used to predict the results of a planned experiment with data close to those from the dataset with a much greater guarantee for the accuracy of the model.

It should be noted that for each of the existing types of lasers, at fixed geometric dimensions, the maximum efficiency is not reached at the maximum values of energy characteristics such as input electric power, pulse repetition frequency, etc. For example, for the considered copper bromide vapor lasers, this is demonstrated by the comparison of the data from the first vmax variant and the next 3 rows of actual data.

5.4.2. Prediction of Laser Output Power

We have to note that the most important characteristic of laser devices is their output power, since it determines their field of application. Efficiency is important but not as much as for other widely used devices.

The next Table 5.6 presents the results of the application of model (5.11)-(5.12) for the prediction of future devices of the type large-bore lasers (D>40mm).

Table 5.6. Variants of the presumed future experiments and predicted values of output power *PredPout* using model (5.11)-(5.12)*

Variant	*D*	*DR*	*L*	*PIN*	*PL*	*PH2*	*PRF*	*PNE*	*C*	*TR*	*Pred Pout*
v-2	58	58	200	4.5	11.25	0.6	16.5	20	1.3	490	113.8
v-1	58	58	200	4.5	11.25	0.6	17.5	20	1.3	490	113.2
v0	58	58	200	5	12.5	0.6	17.5	20	1.3	490	120.8
v1	58	58	200	5.05	12.63	0.6	17.5	20	1.3	490	121.5
v2	60	60	210	5.1	12.14	0.6	17.5	20	1.3	490	125.3
v3	65	63	215	5.15	11.98	0.6	16.5	20	1.3	490	129.2
v4	68	65	220	5.2	11.82	0.6	17.5	20	1.3	500	132.2
v5	68	65	220	5.3	12.05	0.65	18	20	1.1	500	154.8
v6	70	70	240	5.3	11.04	0.65	18	20	1.1	500	165.4
v7	70	70	220	5.3	12.05	0.65	18	20	1.1	500	165.4
v8	70	70	230	5.3	11.52	0.65	18	20	1.1	500	164.4
v9	70	70	230	5.3	11.52	0.65	18	20	1.2	500	151.7
v10	70	70	240	5.3	11.04	0.65	18	20	1.2	500	152.8
v11	70	68	240	5.3	11.04	0.6	17.5	20	1.3	500	138.7
v12	70	70	240	5.5	11.46	0.65	18	20	1.3	500	143.5
v13	70	70	240	5.5	11.46	0.6	17.5	20	1.3	500	143.4
v14	70	68	225	5.25	11.67	0.6	17.5	20	1.3	500	136.0
v15	75	73	230	5.3	11.72	0.6	17.5	20	1.2	500	155.4
v16	75	73	230	5.3	11.72	0.6	17.5	20	1.3	500	141.5

* The first three rows contain data from actual experiments.

5.4.3. Physics Interpretation of the Predictions

The resulting predictions of laser efficiency *Eff* in Table 5.5 indicate that it can be improved by increasing the geometric dimensions *D* and *L* at a constant power supply *PIN*=const. This is confirmed by the results of variants v5 to v10. These variants show that for a fixed *PIN*=5.3 kW, laser efficiency increases from 2.531 to 2.900%. The same results also indicate that the dependence on *L* is more clearly defined. The reason for this is that when the length of the laser tube is increased, the losses along the electrodes are reduced and a larger portion of the electric power is supplied to the active laser volume.

The results from the prediction of laser output power *Pout*, presented in Table 5.6, do not differ significantly in terms of behavior from those for *Eff*. What is new here is the fact that in addition to the geometric dimensions L and *D*, the increase of *Pout* is influenced directly and to a large degree by the supplied electric power *PIN*. This is obvious when all variants given in Table 5.6 are considered.

The experiment reveals that as a whole the quantities PRF, PNE C, TR exert a weaker influence on efficiency and output power and therefore their values are constant or change insignificantly in the selected future experiments.

From a practical point of view, the constructed models are adequate and provide an adequate description of the relationship between input laser characteristics and laser efficiency and output power. To some extent, the models can be used to direct the experiment for the development of new laser devices with improved output laser characteristics.

The comparison between parametric and nonparametric methods for modeling the output power of a copper bromide vapor laser showed that nonparametric models have better general characteristics. The constructed MARS models allow for a more adequate description of the data by overcoming the problems of multicolinearity, local nonlinearities and interactions between predictors. The models can be used both for estimation of known experiments and for the prediction of future ones.

Conclusion

We have obtained the best MARS models of zero order and of first and second order for predicting output laser characteristics of copper bromide lasers. The models were applied for the prediction of an existing and a future experiment. The results indicated an excellent predictive ability of the models.

In comparison with the parametric regression models from Chapter 3 and 4, the constructed nonparametric MARS models are superior both in terms of their high accuracy of approximation and in terms of the stability of the predictions.

REFERENCES

[1] http://www.salfordsystems.com/mars.php.

[2] D. Steinberg, B. Bernstein, P. Colla and K. Martin, *MARS User Guide, Salford Systems*, San Diego, CA, 2001.

[3] K.D. Dimitrov and N.V. Sabotinov, High-power and high-efficiency copper bromide vapor laser, *3052 SPIE* (1996) 126–130.

[4] D. N. Astadjov, K. D. Dimitrov, D. R. Jones, V. K. Kirkov, C. E. Little, N. V. Sabotinov, et al., Copper bromide laser of 120-W average output power, *IEEE J. Quantum Electron.*, (5) 33 (1997) 705–709.

Chapter 6

MODELS OF ULTRAVIOLET COPPER-ION LASERS

ABSTRACT

The goal of this chapter is to construct nonparametric regression models for estimation and prediction of the laser output power (laser generation) of the UV copper ion excited copper bromide vapor laser, described in Chapter 1.

The following problems are solved:

- Classifying the main groups and relationships between input laser variables and determining the influence of the clusters on laser output power
- Constructing and analyzing MARS models, which describe and provide estimates of laser output power
- Applying the constructed MARS models when predicting the laser output power for known and future experiments.

The results in this chapter have been obtained on the basis of the available experimental data for a UV Cu+ Ne-CuBr laser, published in [1-6].

All calculations have been performed with the help of the statistical software packages SPSS [7] and SPM [8] (see also [9, 10]).

Some of the classification results, obtained through cluster analysis have been published in [11]. The results of the modeling with MARS have been published in [12, 13].

6.1. Preliminary Statistical Analysis

The investigated ultraviolet copper ion laser generates simultaneously at five wavelengths – 248.6 nm, 252.9 nm, 259.7 nm, 260.0 nm and 270.3 nm, in a nanosecond pulsed longitudinal Ne-CuBr discharge. A record average output power of 1.3 W was obtained through simultaneous generation at the five ultraviolet lines. An average output power of 0.85 W and peak pulse power of 3.25 W were measured at the 248.6 nm line [2]. It has been established that adding small amounts of hydrogen leads to a twofold increase of laser output power.

The main goal of the future development of this type of laser is to investigate the possibilities for increasing laser output power.

6.1.1. Data Description

A total of 7 laser characteristics (variables) are investigated, which are of physical significance and are considered independent. The variables we will examine are (see also Chapter 2): *D,* mm – inside diameter of the laser tube, *L*, cm – length of the active zone (electrode separation), *PNE*, Torr – neon gas pressure, *PRF*, kHz – pulse repetition frequency, *PIN*, kW – input electrical power, *PH2*, Torr – hydrogen gas pressure, *TR*, °C – temperature of the copper bromide reservoir.

The dependent variable is laser output power *Pout*, mW.

The study uses the data from 251 experiments. All of the data have been published in [1-6].

The basic descriptive statistics of the data on the 7 independent variables and *Pout* are given in Table 6.1. The values of skewness and kurtosis suggest that distributions of some variables are skewed to the left and for others – to the right. This indicates that the variables are not normally distributed. This fact can also be checked using a nonparametric one-sample Kolmogorov-Smirnov test.

Table 6.2 contains correlation coefficients of all variables involved in the analysis. No strong correlations are observed. The determinant of the matrix is sufficiantly different from zero, which also indicates an absence of multicollinearity. For this reason, the considered data are not suitable for parametric regression techniques and cannot be processed in the same way as the data for a copper bromide vapor laser.

Table 6.1. Basic descriptive statistics of the 7 input variables and laser power *Pout* (non-standardized)

	Minimum	Maximum	Mean	Std. Deviation	Variance	Skewness	Kurtosis
	Statistic	Statistic	Statistic	Statistic	Statistic	Statistic	Statistic
D	4.00	26.00	9.61	4.668	21.79	1.854	4.736
L	80.00	86.50	84.52	2.709	7.34	-1.067	-0.842
Pin	1050.00	1900.00	1415.02	211.173	44593.90	1.307	1.222
PNe	7.00	31.88	16.08	4.621	21.35	0.602	1.387
PH2	0.00	0.07	0.006	0.015	0.000	2.572	5.621
PRF	19.50	26.00	21.60	2.454	6.02	0.536	-1.542
Tr	492.00	768.00	570.47	48.731	2374.71	3.701	12.519
Pout	0.14	1300.00	339.17	299.584	89750.62	1.307	1.239
Valid N	251						

Table 6.2. Correlation matrix[a] of 7 input variables and laser output power *Pout* for UV copper ion lasers

		D	L	PIN	PNE	PH2	PRF	TR	Pout
Correlation	D	1	-0.612	-0.482	-0.786	0.078	-0.081	-0.210	-0.493
	L	-0.612	1	0.167	0.580	0.293	-0.338	0.214	0.399
	PIN	-0.482	0.167	1	0.301	-0.119	0.563	-0.033	0.780
	PNE	-0.786	0.580	0.301	1	0.036	0.082	0.067	0.451
	PH2	0.078	0.293	-0.119	0.036	1	-0.342	0.379	0.090
	PRF	-0.081	-0.338	0.563	0.082	-0.342	1	-0.208	0.423
	TR	-0.210	0.214	-0.033	0.067	0.379	-0.208	1	0.077
	Pout	-0.493	0.399	0.780	0.451	0.090	0.423	0.077	1

[a] Determinant = 0.011.

6.2. Classification of the Basic Parameters of Ultraviolet Copper Ion Lasers

The first stage of modeling is to determine the reliable variables of the model and possibly to reduce the number of variables by removing those which do not contribute significantly to the existing relationships. To this end, in this section the 7 basic laser characteristics are classified using agglomerative cluster analysis methods. The influence of the variables on laser generation is also considered.

The investigation was initially conducted using a sample of 136 experiments [11]. Here, CA is performed on all available data with sample size n= 251 experiments.

6.2.1. Construction of Cluster Models

As described in Chapter 2, Cluster analysis is a research technique, which does not place special demands on the data, in particular with regard to their distribution.

In our case, since the sample size is not very large, we apply hierarchical agglomerative methods. The similarity between objects is represented by the most widely used metric – squared Euclidean distance. Since the physical dimensions of the data are different, they are standardized.

The proximity matrix is calculated in the first step of clustering. It is given in Table 6.3. The smallest distance is between *Pout* and *PIN*, equal to 110.21. The next closest variables to *Pout* are *PNE* and *PRF*, respectively at a distance of 274.42 and 288.34.

Table 6.3. Proximity matrix of the 7 input independent variables and Pout with squared Euclidean distance for UV copper ion lasers

	Matrix File Input							
Case	D	L	Pin	PNe	PH2	PRF	Tr	Pout
D	0	806.05	741.06	893.21	461.16	540.68	605.20	746.54
L	806.05	0	416.71	210.18	353.61	668.80	393.05	300.47
Pin	741.06	416.71	0	349.33	559.70	218.69	516.65	110.21
PNe	893.21	210.18	349.33	0	481.99	458.90	466.32	274.42
PH2	461.16	353.61	559.70	481.99	0	671.18	310.25	455.04
PRF	540.68	668.80	218.69	458.90	671.18	0	604.24	288.34
Tr	605.20	393.05	516.65	466.32	310.25	604.24	0	461.42
Pout	746.54	300.47	110.21	274.42	455.04	288.34	461.42	0

Two clustering methods have been applied: Average linkage Between groups and Nearest neighbor, Single Linkage, see formulas (2.15) and (2.16).

For the 7 basis variables, using the Average linkage method and the squared Euclidean distance metric, we obtain the cluster models, given in Table 6.4.a), and using the Single linkage method – the models in Table 6.4.b).

Table 6.4. Cluster models for 7 variables with 2 to 6 clusters obtained using the squared Euclidean distance through the methods: a) Average linkage (between groups) and b) Nearest neighbor

a) Cluster Membership – Average Linkage				
Case	5 Clusters	4 Clusters	3 Clusters	2 Clusters
D	1	1	1	1
L	2	2	2	2
PIN	3	3	3	2
PNE	2	2	2	2
PH2	4	4	2	2
PRF	3	3	3	2
TR	5	4	2	2

b) Cluster Membership – Nearest Neighbor				
Case	5 Clusters	4 Clusters	3 Clusters	2 Clusters
D	1	1	1	1
L	2	2	2	2
PIN	3	3	2	2
PNE	2	2	2	2
PH2	4	4	3	2
PRF	3	3	2	2
TR	5	4	3	2

It is seen that the formal grouping of the variables into 3 clusters is different in the two tables, but for 4 and 5 clusters it is the same. For this reason, the solution with 4 clusters is chosen as the most adequate for our data. These clusters are the following (given according to their proximity and the order of clustering):

$$\begin{aligned} &\text{Cluster 1:} \quad \{PNE, L\} \\ &\text{Cluster 2:} \quad \{PIN, PRF\} \\ &\text{Cluster 3:} \quad \{PH2, TR\} \\ &\text{Cluster 4:} \quad \{D\} \end{aligned} \tag{6.1}$$

The graphic representation of the cluster model obtained from Average linkage method is shown using the dendrogram in Figure 6.1. The added vertical dotted lines to the figure indicate that due to the significant jumps of *PH2* and *TR*, the two should be separated into different clusters, with *TR* being close to the boundary jump of 5 measurement units.

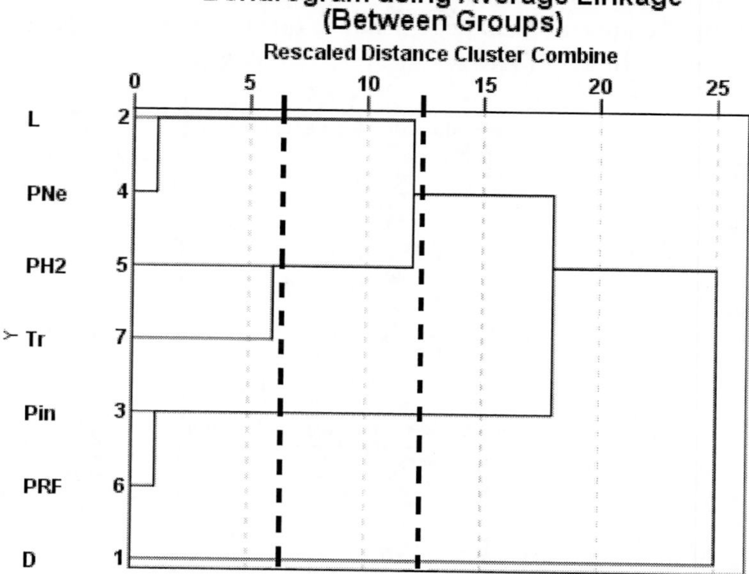

Figure 6.1. Dendrogram of the 7 variables using hierarchical cluster analysis, method of Average linkage, for UV copper ion laser.

In order to determine the influence of the obtained clusters on laser output power, we repeat the previously described procedures for all data (n=251), also including *Pout*. The proximity coefficients at the first stage of clustering are shown in Table 6.3. From the respective dendrogram in Figure 6.2, with the help of the Average linkage (Between groups) method, almost the same solution as (6.1) is obtained with a 5 cluster structure of the links, similarly to Figure 6.1, with some minor shifts. It is apparent that the second cluster {*PIN*, *PRF*} includes *Pout*, and therefore has the strongest impact on *Pout*. To be more precise, the distance between *PIN* and *Pout* is the smallest, it is 110.21. The second closest to *Pout* cluster is {*PNE*, *L*}. The next clusters are {*PH2*, *TR*}, and {*D*}. The solution with other clustering methods, which have not been presented here, is analogical.

The exact influence of individual clusters on output power will be investigated using suitable multivariate regression methods.

It can be concluded that the obtained cluster models can be used as a reference point when planning the experiment aimed at increasing output power. This is especially helpful having in mind that the geometric design of existing laser devices cannot be changed. In this case, the cluster {*D*} cannot be changed. The cluster {*PNE L*} can be modified to some extent, since the length of the active zone *L* is preset for each laser. It was established from Table 6.3 that the cluster {*PIN*, *PRF*} has the strongest influence on output power. To be more specific, an increase of *PIN* leads to a direct increase of *Pout* within certain limits. The constructed models also indicate that the increase of *Pout* for new laser devices can be achieved by increasing *PIN* and *L*, as well as by possible reducing of *D*. These findings correspond well with experimental studies [1-6].

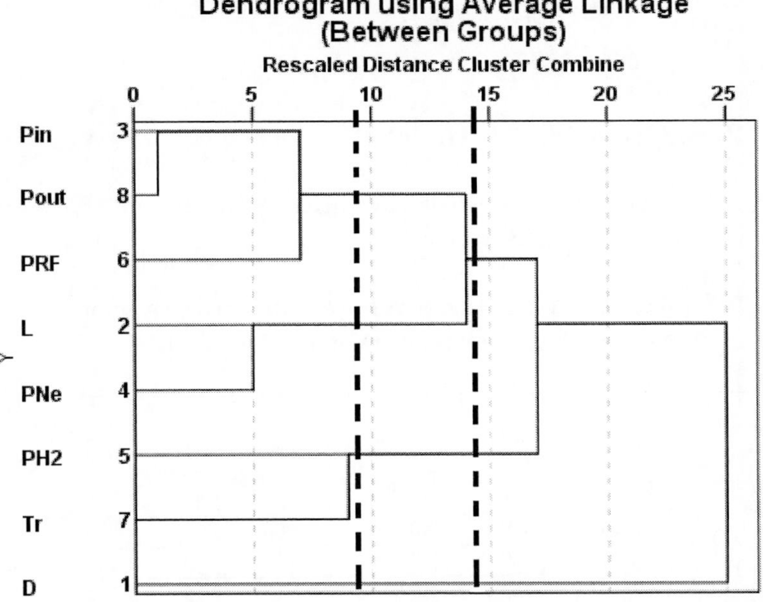

Figure 6.2. Dendrogram of the 7 variables and *Pout* using Average linkage for UV copper ion laser.

6.3. MARS Models of the Laser Generation of Ultraviolet Copper Ion Lasers

In this section, flexible regression models describing laser generation *Pout* are constructed and examined with the help of the nonparametric MARS technique.

The study is conducted using the data from n=251 experiments, described in 6.1. It was not established that their distribution is normal or close to normal. However, FA was formally performed, which showed adequacy with distinguishable factor variables. Nevertheless, the regression models did not demonstrate satisfactory predictive qualities. For this reason, nonparametric methods have been used, and more specifically - MARS models.

After constructing the models, it was established that the following laser variables (predictors) have influence: *D, L*, *PNE*, *PIN*, *PRF*, *PH2*, and *TR*. The dependent variable is *Pout* – laser output power.

All calculations have been performed using the SPM MARS statistical software [8].

6.3.1. Zero Order MARS Models

The first type of models, which have been constructed are MARS models without interaction between predictors.

Table 6.5. Statistical indices of the constructed MARS models of output laser power of UV copper bromide lasers

Model type	Max BF	R2	R2 adj.	R2 GCV	BF in the model	Predic tors
0th order	15	0.918	0.914	0.900	10	5
	20	0.929	0.925	0.908	12	5
	25	**0.937**	**0.934**	**0.917**	**13**	**5**
	30	0.938	0.935	0.918	14	5
1st order	15	0.954	0.952	0.939	12	5
	20	**0.967**	**0.965**	**0.954**	**15**	**5**
	25	0.975	0.973	0.960	19	6
	30	0.977	0.975	0.962	19	6
	35	**0.982**	**0.980**	**0.966**	**24**	**6**
	40	0.982	0.980	0.966	24	6
2nd order	15	0.955	0.953	0.940	13	5
	20	0.966	0.964	0.952	15	5
	25	0.973	0.971	0.958	18	5
	30	0.976	0.974	0.960	20	6
	35	**0.980**	**0.978**	**0.963**	**25**	**5**
	40	0.982	0.980	0.967	26	6

The basis indices of four such zero order models are given in Table 6.5. The last two of these have the highest and almost similar indices, and therefore, here we will present the results of the model with a maximum of 25 BF.

In the case with a maximum of 25 basis functions, the zero order best MARS model includes the following 13 BF:

$$\begin{aligned}
&BF1 = \max(0, PIN - 1440)\\
&BF3 = \max(0,\ PNE - 19.34)\\
&BF4 = \max(0, 19.34 - PNE)\\
&BF5 = \max(0,\ PH2 - 0.03)\\
&BF6 = \max(0, 0.03 - PH2)\\
&BF9 = \max(0, PNE - 20)\\
&BF11 = \max(0, PNE - 18.75)\\
&BF15 = \max(0, PIN - 1400)\\
&BF17 = \max(0, D - 5.7)\\
&BF19 = \max(0, PRF - 19.5)\\
&BF20 = \max(0, PIN - 1600)\\
&BF22 = \max(0, PIN - 1340)\\
&BF24 = \max(0, D - 7.1)
\end{aligned} \tag{6.2}$$

The regression model for laser output power *Pout* with these functions is:

$$\begin{aligned}
Pout = {} & 845.548 + 1.537\ \mathrm{BF1} - 947.68\ \mathrm{BF3} - 9.723\ \mathrm{BF4}\\
& - 126\ \mathrm{BF5} - 15309.2\ \mathrm{BF6} + 477.71\ \mathrm{BF9} + 452.01\ \mathrm{BF11}\\
& - 3.85245\ \mathrm{BF15} - 170.98\ \mathrm{BF17} + 13.514\ \mathrm{BF19}\\
& + 1.436\ \mathrm{BF20} + 2.132\ \mathrm{BF22} + 167.54\ \mathrm{BF24}
\end{aligned} \tag{6.3}$$

In (6.3) the reliable predictors are five: PIN, $PH2$, PNE, D, PRF.

The respective importance of the predictors within the model are given in the general Table 6.6.

Figs. 6.3 illustrate the relationship between predictor variables and laser output power *Pout*. It is apparent that *Pout* reaches a maximum when *PH2*=0.03, peaks when *PNE* is roughly 19-20 units and increases together with *PIN*.

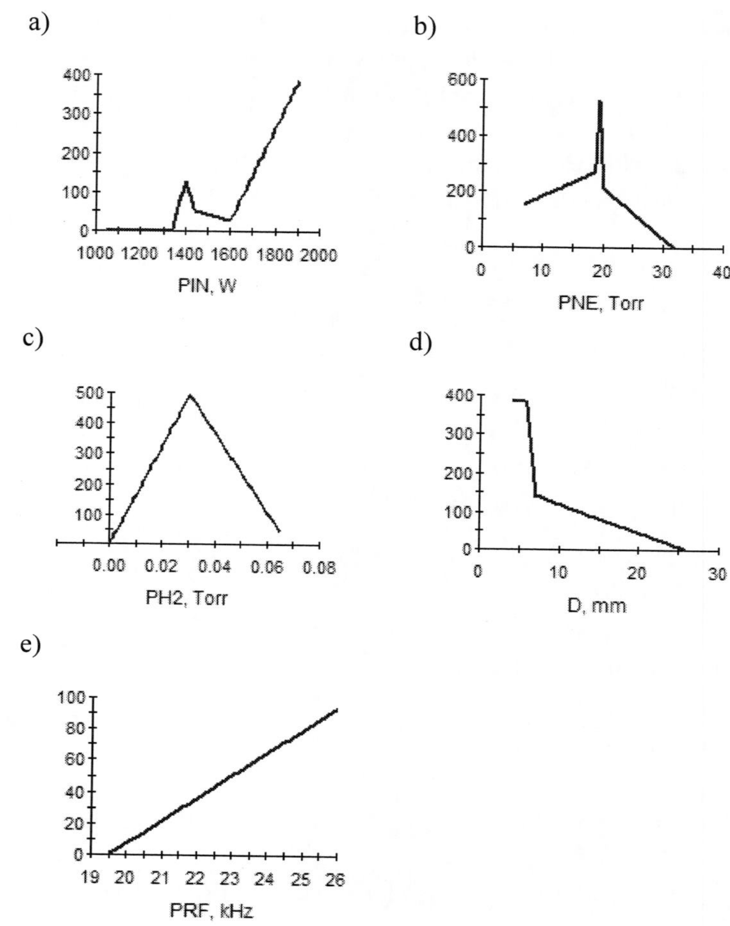

Figure 6.3. Influence of each predictor from model (6.2)-(6.3) on *Pout*, in pure ordinal units.

With the help of the model (6.2)-(6.3), the value of *Pout* is easily calculated when the predictors are known. For example, the maximum measured laser power *Pout* = 1300 mW is achieved for the following values of the laser input characteristics: *D*=5.7 mm, *PIN*=1900 W, *PNE*=19.3 Torr, *PH2*=0 Torr, *TR*=560 K, *L*=86 cm и *PRF*=25 KHz [2]. After substituting these in (6.2)-6.3) we get the approximate estimate of $\hat{P}out = 1114$ mW. This result is quite unsatisfactory and for this reason, we construct models with interaction between basis functions.

Table 6.6. Relative variable importance in presented MARS models of the output power in UV copper ion laser

Variable	Model (6.2)-(6.3)	Model (6.4)-(6.5)	Model (6.6)-(6.7)	Model (6.8)-(6.9)
PIN	100.00	100.00	100.00	100.00
PNE	77.18	82.30	99.89	78.32
PH2	82.81	48.14	58.76	53.07
D	41.82	47.36	55.48	87.87
PRF	21.54	18.44	25.56	22.30
TR	0.0	0.0	20.70	0.0
L	0.0	0.0	0.0	0.0

6.3.2. First Order Best MARS Models

Analogically to the previous Chapter 5, we will also construct the best MARS models with possible first order interactions between the predictors. We expect these will provide better approximations of the actual behavior of laser output power.

Six models are presented in the case of possible first order interaction between the predictors. Their statistical indices are given in Table 6.5.

We will first present the first order model with up to 20 BF. It is characterized by some of the highest indices of all models with five predictors. The model employs the following 15 BF:

$$
\begin{aligned}
&BF1 = \max(0, PIN - 1440)\\
&BF2 = \max(1440 - PIN)\\
&BF3 = \max(0,\ PNE - 19.34)\\
&BF4 = \max(0, 19.34 - PNE)\\
&BF5 = \max(0,\ PH2 - 0.03)\ BF4\\
&BF6 = \max(0,\ 0.03 - PH2)\ BF4\\
&BF8 = \max(0,\ 19.3 - PNE)\ BF1\\
&BF9 = \max(0,\ D - 8)\, BF2\\
&BF10 = \max(0, 8 - D)\, BF2\\
&BF11 = \max(0, PIN - 1600)\, BF3\\
&BF13 = \max(0, PNE - 25.625)
\end{aligned}
\qquad (6.4)
$$

$$BF15 = \max(0, PIN - 1400)$$
$$BF18 = \max(0,\ 20 - PRF)$$
$$BF19 = \max(0,\ D - 9.5)\,BF18$$
$$BF20 = \max(0,\ 9.5 - D)\,BF18$$

The regression equation of the model is

$$\begin{aligned} \hat{P}out = {} & 377.12 + 4.451\ BF1 - 1.384\ BF2 - 99.285\ BF3 \\ & + 143.6\ BF4 - 4405.72\ BF5 - 5250.25\ BF6 - 1.48\ BF8 \\ & + 0.048\ BF9 + 1.097\ BF10 - 2.394\ BF11 + 104.29\ BF13 \\ & - 2.342\ BF15 - 57.07\ BF19 - 63.409\ BF20 \end{aligned} \qquad (6.5)$$

The relative importance of the predictors PIN, PNE, $PH2$, D, PRF in the model is given in the general Table 6.5, first order interaction models.

Figure 6.4 displays the relationship between included predictor variables and their interactions on laser output power *Pout*.

a)

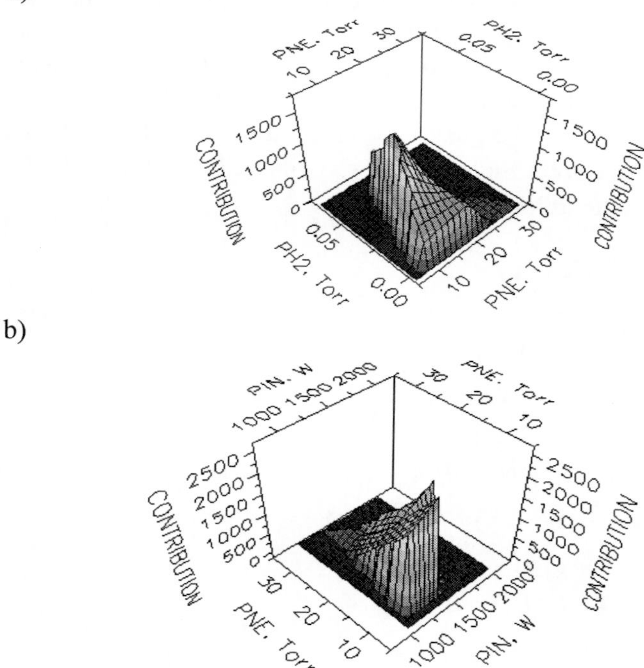

Figure 6.4. (Continued).

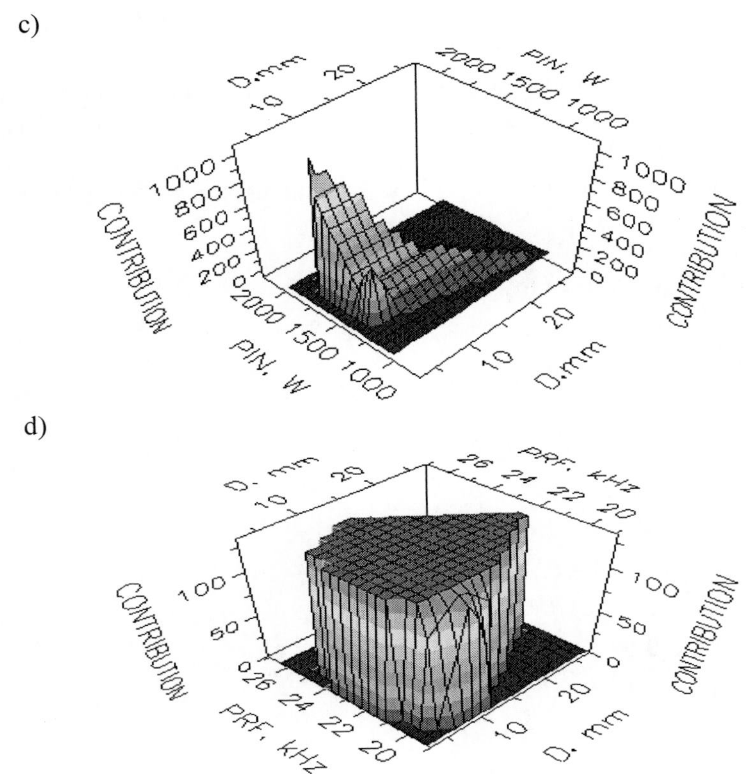

Figure 6.4. Influence of the first order interactions between the predictors in model (6.4)-(6.5) on *Pout* of UV lasers, in pure ordinal units.

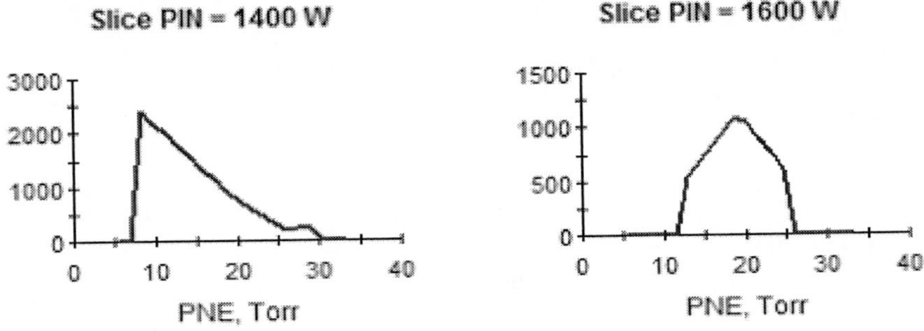

Figure 6.5. View of two slices from Figure 6.4.a: the first shows an alteration of the influence of *PNE* on *Pout* at a fixed *PIN*=1400 W, and the second – at a fixed *PIN*=1600 W in model (6.4)-(6.5) in pure ordinal units.

Figure 6.5 shows in slices the details of the interaction between *PIN* and *PNE*.

Of the first order models, those with the highest (and equal) indices are the last two models, with up to 35 and 40 BF (see Table 6.5) respectively. They use 6 predictor variables.

In the case with a maximum of 35 basis functions, the model includes the following 27 BF:

$$\begin{aligned}
&BF1 = \max(0, PIN - 1440)\\
&BF2 = \max(1440 - PIN)\\
&BF3 = \max(0,\ PNE - 19.34)\\
&BF4 = \max(0, 19.34 - PNE)\\
&BF5 = \max(0,\ PH2 - 0.03)\ BF4\\
&BF6 = \max(0,\ 0.03 - PH2)\ BF4\\
&BF7 = \max(0,\ PNE - 19.3)\ BF1\\
&BF8 = \max(0,\ 19.3 - PNE)\ BF1\\
&BF9 = \max(0,\ D - 8)\, BF2\\
&BF10 = \max(0, 8 - D)\, BF2\\
&BF11 = \max(0, PIN - 1600)\, BF3\\
&BF13 = \max(0, PNE - 25.625)\\
&BF14 = \max(0,\ 25.625 - PNE)\\
&BF15 = \max(0, PIN - 1400)\\
&BF18 = \max(0,\ 20 - PRF)\\
&BF19 = \max(0,\ D - 9.5)\, BF18\\
&BF20 = \max(0,\ 9.5 - D)\, BF18\\
&BF23 = \max(0,\ TR - 492)\, BF2\\
&BF24 = \max(0,\ PRF - 20)\, BF4\\
&BF25 = \max(0,\ 20 - PRF)\, BF4\\
&BF27 = \max(0,\ 1600 - PIN)\, BF14\\
&BF28 = \max(0,\ TR - 560)\, BF4\\
&BF29 = \max(0,\ 560 - TR)\, BF4\\
&BF30 = \max(0,\ D - 12)
\end{aligned} \qquad (6.6)$$

$$BF31 = \max(0, 12 - D)$$
$$BF32 = \max(0, \ PNE - 19.85)\,BF15$$
$$BF34 = \max(0, \ PNE - 26.25)\,BF2$$

Corresponding regression equation for predicting output power is

$$\begin{aligned} \hat{P}out = {} & 375.64 + 4.596\ BF1 - 100.52\ BF3 + 124.81\ BF4 \\ & - 4551.02\ BF5 - 5846.81\ BF6 + 3.198\ BF7 - 1.33\ BF8 \\ & - 0.2727\ BF9 + 1.4625\ BF10 - 6.9594\ BF11 \\ & + 330.225\ BF13 - 1.971 BF15 - 107.266\ BF19 \\ & - 58.2\ BF20 - 0.0122\ BF23 + 4.32978\ BF24 \\ & + 53.491\ BF25 + 0.0551\ BF27 + 0.3379\ BF28 \\ & - 0.3934\ BF29 + 130.016\ BF30 - 45.117\ BF31 \\ & - 2.701\ BF32 - 1.708\ BF34 \end{aligned} \tag{6.7}$$

The included predictors in (6.6) - (6.7) are six: *PIN*, *PNE*, *PH2*, *D*, *PRF*, *TR*. It is immediately verified that (6.6) includes all BF from (6.4). The result is that model (6.6)-(6.7) with up to 35 BF can be viewed as a clarification and an improvement of model (6.4)-(6.5) with up to 20 BF.

The respective inportance of these predictors in (6.7) is given in Table 6.6. Plots of the contributions of the interacting members are illustrated in Figure 6.6.

a)

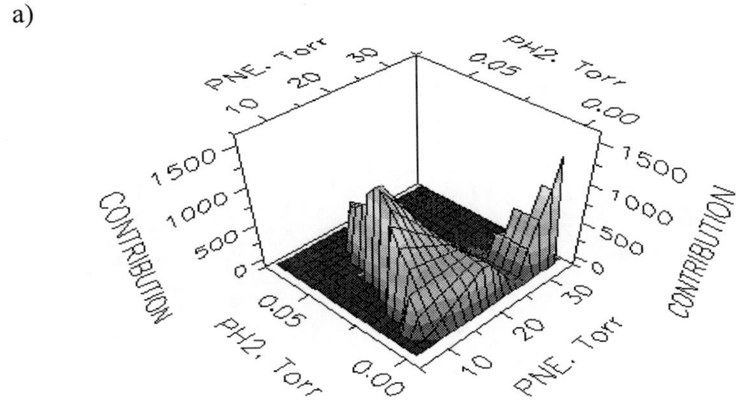

Figure 6.6. (Continued).

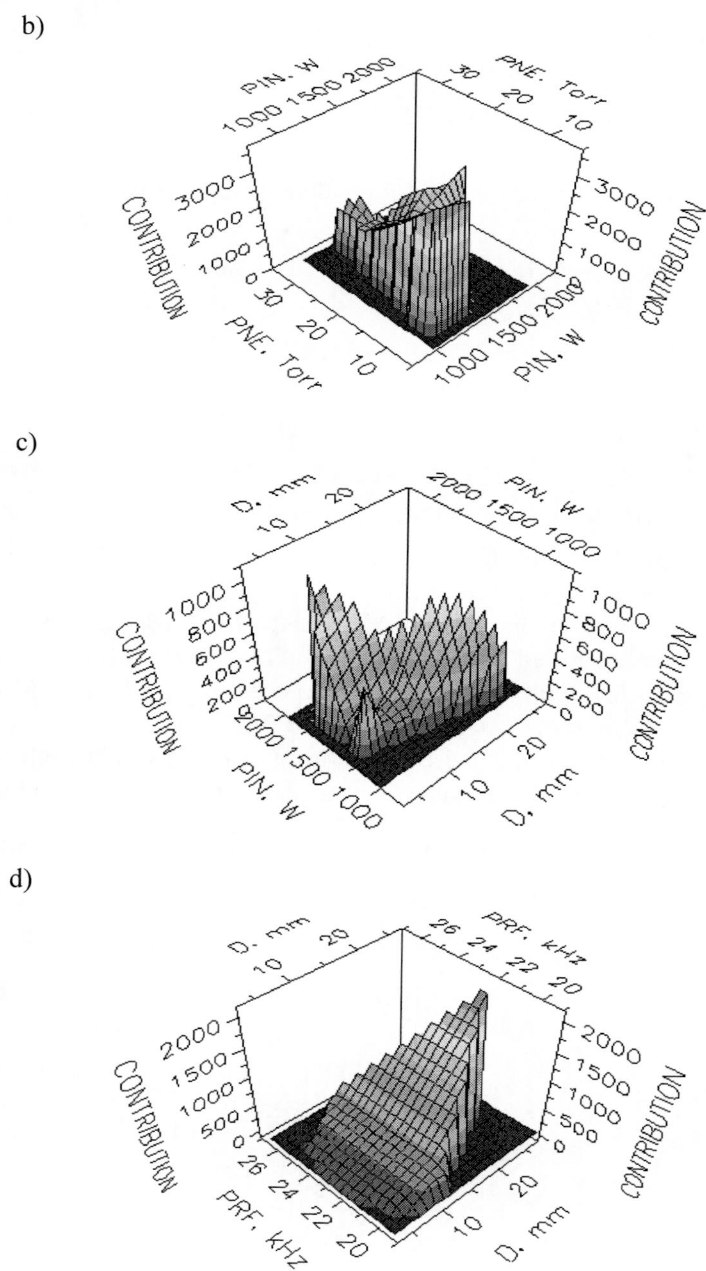

Figure 6.6. (Continued).

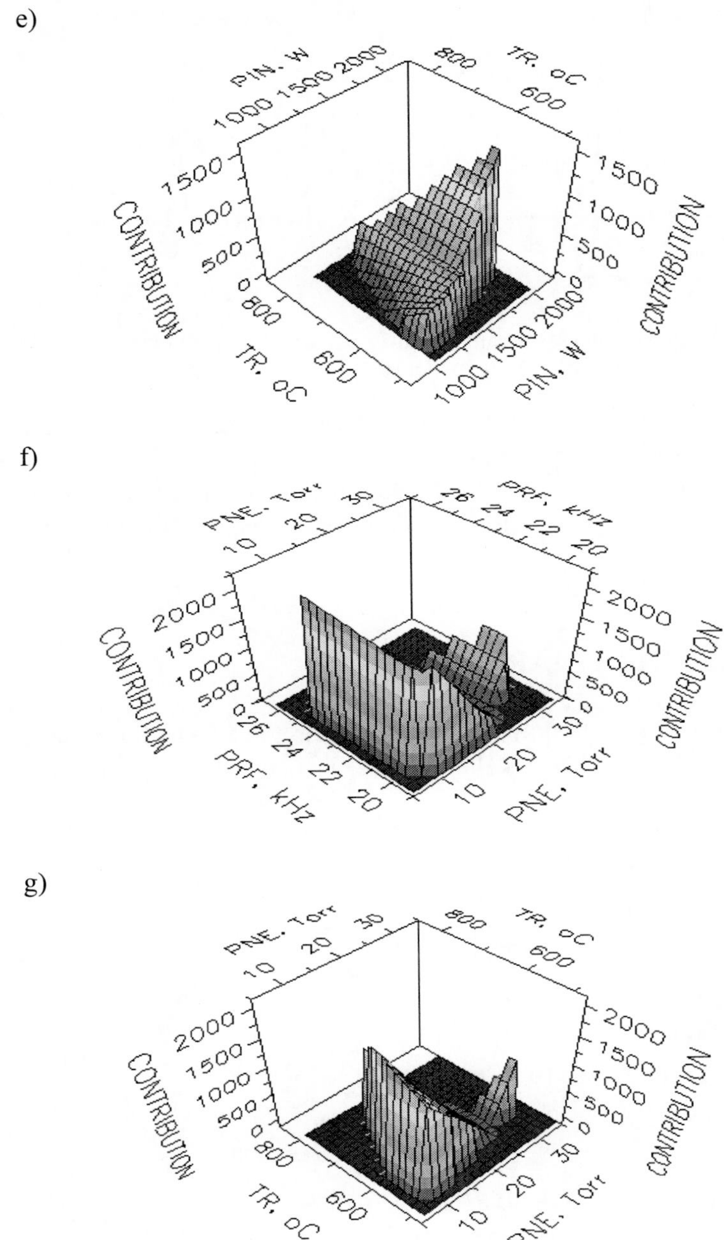

Figure 6.6. Influence of the first order interactions between the predictors in model (6.6-(6.7) on *Pout* in pure ordinal units.

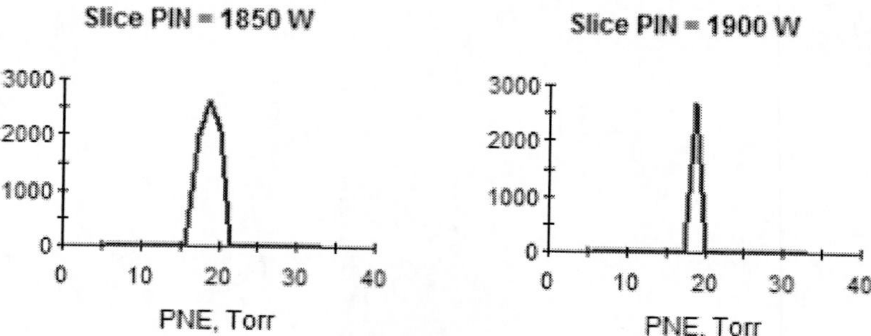

Figure 6.7. Example of slices, showing the influence of *PNE* on *Pout* in pure ordinal units when *PIN*=1850 and 1900 W in model (6.6)-(6.7).

As an example, Figure 6.7 shows two slices from the modeled behavior of the variable *PNE* (neon gas pressure) for two fixed values of input power *PIN*=1850 W and *1900* W, obtained using model (6.6)-(6.7), as elements of Figure 6.6b. In this case, there is a clearly identifiable maximum in the interval 18-19 Torr.

6.3.3. Second Order MARS Models

Here, we will present only the model with up to 35 basis functions and 5 predictors. The best MARS model in this case, includes the following 25 BF:

$$\begin{aligned}
&BF1 = \max(0, PIN - 1440) \\
&BF2 = \max(1440 - PIN) \\
&BF3 = \max(0,\ PNE - 19.34) \\
&BF4 = \max(0, 19.34 - PNE) \\
&BF5 = \max(0,\ PH2 - 0.03)\ BF4 \\
&BF6 = \max(0,\ 0.03 - PH2)\ BF4 \\
&BF7 = \max(0,\ PNE - 19.3)\ BF1 \\
&BF8 = \max(0,\ 19.3 - PNE)\ BF1 \\
&BF10 = \max(0, 8 - D)\, BF2 \\
&BF11 = \max(0, PIN - 1600)\, BF3 \\
&BF13 = \max(0, PNE - 21.875)\, BF10 \\
&BF14 = \max(0,\ 21.875 - PNE)\, BF10
\end{aligned} \tag{6.8}$$

$$BF17 = \max(0, PIN - 1400)\,BF6$$
$$BF19 = \max(0, PRF - 20)\,BF6$$
$$BF23 = \max(0, D - 12)\,BF6$$
$$BF25 = \max(0, PRF - 20)\,BF10$$
$$BF26 = \max(0, 20 - PRF)\,BF10$$
$$BF27 = \max(0, D - 4)\,BF1$$
$$BF28 = \max(0, PNE - 19.85)\,BF27$$
$$BF30 = \max(0, D - 9.5)\,BF2$$
$$BF31 = \max(0, 9.5 - D)\,BF2$$
$$BF32 = \max(0, PRF - 21)\,BF31$$
$$BF33 = \max(0, 21 - PRF)\,BF31$$
$$BF34 = \max(0, PRF - 21)\,BF2$$
$$BF35 = \max(0, 21 - PRF)\,BF2$$

The regression equation for the prediction of laser output power *Pout* with 23 of these functions is in the following form:

$$\begin{aligned}\hat{P}out = {} & 277.831 - 2.3916\,BF2 - 218.645\,BF3 + 162.252\,BF4 \\ & - 4503.71\,BF5 - 5648.8\,BF6 + 2.74\,BF7 - 0.963\,BF8 \\ & + 2.429\,BF10 - 5.811\,BF11 + 0.405\,BF13 - 0.239\,BF14 \\ & - 13.395\,BF17 + 89.506\,BF19 + 468.97\,BF23 - \\ & - 1.1066\,BF25 - 0.888\,BF26 + 1.210\,BF27 \\ & - 1.8169\,BF28 - 0.2994\,BF30 + 0.7388\,BF32 \\ & - 0.2229\,BF33 + 0.4489\,BF34 + 1.19\,BF35\end{aligned} \tag{6.9}$$

The five predictors involved in the model are: *PIN, D, PNE, PH2, PRF*. Their respective importance in the model is given in Table 6.6.

6.3.4. Comparison of Models and Interpretation

In addition to the abovementioned MARS models, numerous other models, which displayed similar results, were constructed and examined.

It has to be noted that out of the 7 variables, included in the investigation, 5 demonstrate a more significant influence on laser generation. Tables 6.5 and

6.6 show that the strongest influence on *Pout* is that of the following, in descending order: *PIN*, *PNE*, *D*, *PH2* and *PRF*. The variables *L* and *TR* make no significant contribution. However, this fact for the small contribution of *L* and *TR* was not clearly established by the clustering procedures. It needs to be mentioned that the attempts to include other variables, such as *PL* = *PIN* / *L* showed that these are not significant and were thus excluded from the analysis.

The basic statistical indices of some of the investigated models have been systematized in Table 6.5. The chosen best models in this table are given in bold.

The comparison based on results in Table 6.5 indicates that the models without interaction, i.e. those with linear regression splines, as a whole, have worse characteristics than the other types of models. However, these models are easier to interpret, since the graphs of the included variables and the general trends can be monitored. Figure 6.3, as well as the formulas of the models without interaction, explain the behavior of *Pout* when the variables *PIN, PNE, PH2* change, which is almost the same for all obtained models.

The models with first and second order interactions have similar coefficients of determination, although the models with second order interaction perform slightly better, considering that for latter, the number of BF is smaller, the number of relevant variables is higher and the overall quality of the fitting to the data is better. Another important criterion is also the calculated R2-GCV (general cross-validation coefficient of determination), which is more accurate than the standard coefficients of "naïve" determination R2 and R2-adj. The models, obtained for an initial limit of up to 30, 40 and more BF, although characterized by slightly better formal indices, in reality offer worse predictions, particularly for a maximum value of *Pout*, equal to 1300 mW. This is explained by the fact that these models demonstrate a certain overfitting, i.e. the measurement error is comparable with model accuracy. The chosen best models, which fit the listed criteria, are given in bold in Table 6.8.

The interpretation of the nonparametric models can be presented in detail by considering individual subsets, within which the local variation of the dependent variable is of interest.

It can be concluded that the relationship between laser input variables and Pout is characterized by local second and third order nonlinearities in individual subsets of the investigated multidimensional data set and the constructed models fit the data very well.

6.3.5. Prediction of Laser Generation of UV Copper Ion Lasers

In this Chapter, we have shown that the constructed models have very good predictive qualities with regard to the available data. The obtained MARS models can also be used to predict future experiments. In order to illustrate the predictive ability of model (6.6)-(6.7), we will fix the values of laser characteristics: *D*=5.7 mm and *PH*2=0.0 Torr. In this way, a certain subset of the known data is extracted.

When the input power is *PIN*=1900 W and the pressure of the buffer gas neon (the variable *PNE*) is changed in the interval [19.1, 19.6], we obtain the values predicted by model (6.6)-(6.7). They can be compared with the data from the experiments. The results are illustrated in Figure 6.8. Their numerical values are given in Table 6.7.

It can be noted that the deviation is more significant towards the boundaries of the interval. This is not incidental, since within the considered range, the maximum for *PNE*=19.3 has been established through experiments, and for lower or higher values the output power decreases. Here, the trend in the calculations is stronger than that observed from the measurements. The deviation error in the middle of the interval (the maximum), both for the measurement and the model, is within 5-6%.

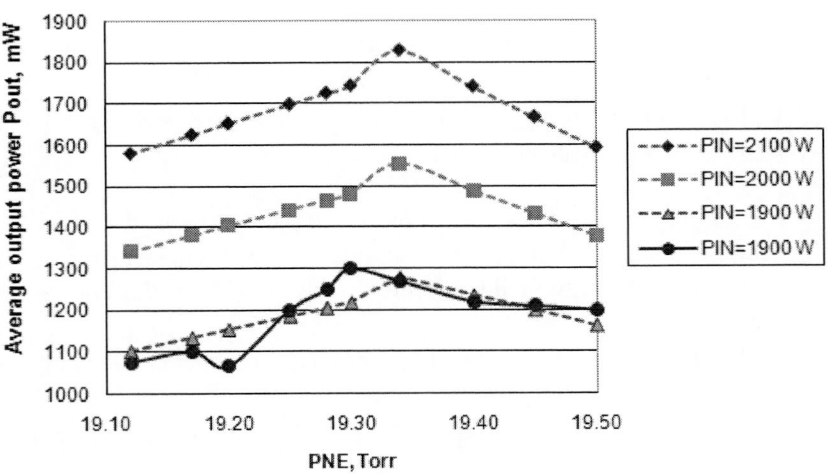

Figure 6.8. Comparison between experimental data for laser power *Pout* (solid line) and the values predicted by model (6.6)-(6.7) (squares with dashed line) when *PIN*=1900; the other two lines represent predicted future experiments when *PIN*=2000 W and *PIN*=2100 W.

The same Figure 6.8 also shows the calculated predicted values for a future experiment if the value of the input power is fixed first at *PIN*=2000 W and then at *PIN*=2100 W. It is observed that the behavior of the output power remains unchanged and it increases by 200 mW and 400 mW, respectively, when compared with experimental data.

Table 6.7. Values of output laser power Pout for different input laser power PIN

PNE, Torr	PIN=2100 W	PIN=2000 W	PIN=1900 W	PIN=1900 W-experiment
19.12	1580	1342	1103	1075
19.17	1626	1381	1135	1100
19.20	1653	1404	1155	1065
19.25	1698	1443	1187	1200
19.28	1726	1466	1206	1250
19.30	1744	1481	1219	1300
19.34	1829	1554	1279	1270
19.40	1741	1488	1236	1220
19.45	1668	1434	1200	1210
19.50	1594	1379	1164	1200

CONCLUSION

The basic results from the statistical modeling of the data for the characteristics of a UV copper ion laser are:

- Cluster models, which determine the basic groups and relationships between the 7 input laser variables have been constructed; the classification of the independent variables into four clusters has been established
- The influence of the clusters on laser output power has been established
- The best MARS models of 0th, 1st, and 2nd order interactions between the predictors and laser generation have been constructed
- The models have been compared, based on the general statistical indices

- It has been established that models of first and second order interactions (with piecewise second and third order relationships) demonstrate the best qualities
- As a whole, the laser characteristics with the highest influence in the regression models correspond with the cluster models.
- The models have been used to predict known and future experiments.

The results of the statistical modeling of the multiline average output power of an ultraviolet copper ion excited copper bromide vapor laser have been obtained using nonparametric MARS models. The models demonstrate good predictive abilities when predicting existing and future experiments. It has been established which laser characteristics are of significance for the increase of laser output power and to what degree.

This approach could be applied successfully to other laser devices of similar type.

REFERENCES

[1] N. K. Vuchkov, UV copper ion laser in Ne-CuBr pulse-longitudinal discharge, In *Advances in Laser and Optics Research*, Ed. W.T. Arkin, Nova Science Publishers Inc., New York, 2002, 1-33.

[2] N. Vuchkov, High discharge tube resource of the UV Cu+ Ne-CuBr laser and some applications, In: *New development in lasers and electric-optics research*, ed. W. T. Arkin, Nova Science Publishers, New York, 2006, 41-74.

[3] N. K. Vuchkov, K. A. Temelkov and N. V. Sabotinov, UV Lasing on Cu+ in a Ne-CuBr Pulsed Longitudinal Discharge, *IEEE J. Quantum Electron.* 35(12) (1999) 1799-1804.

[4] N. K. Vuchkov, K. A. Temelkov, P. V. Zahariev and N. V. Sabotinov, Influence of the active zone diameter on the UV-ion Ne-CuBr laser performance, *IEEE J. Quantum Electron.*, 37(12) (2001) 1538-1546.

[5] N. K. Vuchkov, K. A. Temelkov, P. V. Zahariev and N. V. Sabotinov, Output parameters and a spectral study of UV Cu+ Ne-CuBr laser, *Optics and Laser Technology*, 36(1) (2004) 19-25.

[6] N. K. Vuchkov, K. A. Temelkov and N. V. Sabotinov, Effect of hydrogen on the average output of the UV Cu+ Ne-CuBr laser, *IEEE J. Quantum Electron.,* 41(1) (2005) 62-65.

[7] http://www.spss.com/software statistics/ stats-pro/, IBM® SPSS® Statistics, 2011.

[8] http://www.salfordsystems.com/mars.php, 2011.

[9] J. H. Friedman, Multivariate adaptive regression splines (with discussion), *The Annals of Statistics*, 19(1) (1991) 1-141.

[10] T. J. Hastie and R. J. Tibshirani, *Generalized additive models*, Chapman and Hall/CRC, 1990.

[11] S.G. Gocheva-Ilieva, I.P. Iliev, K. A. Temelkov, N. K. Vuchkov, N. V. Sabotinov, Classifying the basic parameters of ultraviolet Copper Bromide laser, *Int. Conf. AMiTaNS2009, 22-27 June 2009, Sozopol, Bulgaria, AIP Conf. Proc.* eds. M. D. Todorov and C. I. Christov, Melville NY: American Institute of Physics, CP1186 (2009) 413-420.

[12] S. G. Gocheva-Ilieva, Application of MARS for the construction of nonparametric models, *Proceedings of the 39th Spring Conference of the Union of Bulgarian Mathematicians*, Albena, April 6-10, 2010, ed. Peter Russev, (2010) 29-38.

[13] S. G. Gocheva-Ilieva and I. P. Iliev, Modeling and prediction of laser generation in UV copper bromide laser via MARS, in: Advanced research in physics and engineering, series "Mathematics and Computers in Science and Engineering", ed. O. Martin et al., *Proc. of 5th Int. Conf. on Optics, Astrophysics and Astronomy (ICOAA '10)*, Cambridge, UK, February 20-22, 2010, WSEAS Press (2010) 166-171.

INDEX

A

B

C

D

E

F

G

H

I

K

L

M

N

O

P

Q

R

S

T

U

V

W

Z